Dynamic Stability and Control of Tripped and Untripped Vehicle Rollover

Synthesis Lectures on Advances in Automotive Technology

Editor
Amir Khajepour, *University of Waterloo*

The automotive industry has entered a transformational period that will see an unprecedented evolution in the technological capabilities of vehicles. Significant advances in new manufacturing techniques, low-cost sensors, high processing power, and ubiquitous real-time access to information mean that vehicles are rapidly changing and growing in complexity. These new technologies—including the inevitable evolution toward autonomous vehicles—will ultimately deliver substantial benefits to drivers, passengers, and the environment. Synthesis Lectures on Advances in Automotive Technology Series is intended to introduce such new transformational technologies in the automotive industry to its readers.

Electrification of Heavy-Duty Construction Vehicles
Hong Wang, Yanjun Huang, Amir Khajepour, and Chuan Hu
2017

Vehicle Suspension System Technology and Design
Avesta Goodarzi and Amir Khajepour
2017

Dynamic Stability and Control of Tripped and Untripped Vehicle Rollover
Zhilin Jin, Bin Li, and Jingxuan Li

ISBN: 978-3-031-00372-1   paperback
ISBN: 978-3-031-01500-7   ebook
ISBN: 978-3-031-00005-8   hardcover

DOI 10.1007/978-3-031-01500-7

A Publication in the Springer series
SYNTHESIS LECTURES ON ADVANCES IN AUTOMOTIVE TECHNOLOGY

Lecture #6
Series Editor: Amir Khajepour, *University of Waterloo*
Series ISSN
Print 2576-8107   Electronic 2576-8131

Dynamic Stability and Control of Tripped and Untripped Vehicle Rollover

Zhilin Jin
Nanjing University of Aeronautics & Astronautics

Bin Li
Aptiv PLC

Jingxuan Li
Nanjing University of Aeronautics & Astronautics

*SYNTHESIS LECTURES ON ADVANCES IN AUTOMOTIVE TECHNOLOGY
#6*

ABSTRACT

Vehicle rollover accidents have been a serious safety problem for the last three decades. Although rollovers are a small percentage of all traffic accidents, they do account for a large proportion of severe and fatal injuries. Specifically, some large passenger vehicles, such as large vans, pickup trucks, and sport utility vehicles, are more prone to rollover accidents with a high center of gravity (CG) and narrow track width. Vehicle rollover accidents may be grouped into two categories: tripped and untripped rollovers. A tripped rollover commonly occurs when a vehicle skids and digs its tires into soft soil or hits a tripping mechanism such as a curb with a sufficiently large lateral velocity. On the other hand, the untripped rollover is induced by extreme maneuvers during critical driving situations, such as excessive speed during cornering, obstacle avoidance, and severe lane change maneuver. In these situations, the forces at the tire-road contact point are large enough to cause the vehicle to roll over. Furthermore, vehicle rollover may occur due to external disturbances such as side-wind and steering excitation. Therefore, it is necessary to investigate the dynamic stability and control of tripped and untripped vehicle rollover so as to avoid vehicle rollover accidents.

In this book, different dynamic models are used to describe the vehicle rollover under both untripped and special tripped situations. From the vehicle dynamics theory, rollover indices are deduced, and the dynamic stabilities of vehicle rollover are analyzed. In addition, some active control strategies are discussed to improve the anti-rollover performance of the vehicle.

KEYWORDS

vehicle rollover, dynamic stability, tripped rollover, rollover index, rollover warning, anti-roll control, active safety of vehicle

Contents

CHAPTER 1

Introduction

1.1 WHAT IS VEHICLE ROLLOVER?

Vehicle rollover is a dangerous lateral movement which refers to the vehicle in the process of driving around its longitudinal axis rotate 90° or more, so that the body is in contact with the ground.

1.2 RISK OF VEHICLE ROLLOVER ACCIDENTS

Vehicle rollover accidents have been a serious safety problem for the last three decades. Athough rollovers are a small percentage of all traffic accidents, they do account for a large proportion of severe and fatal injuries. According to National Highway Traffic Safety Administration (NHTSA) in the U.S., rollover accidents are the second most dangerous form of accident in the U.S., after head-on collisions. In 2012, almost 5.615 million vehicles crashed in the U.S., although only 2.0% involved vehicle rollover, and the proportion of vehicles that rolled over in fatal crashes was 20.3% [1]. Specifically, some large passenger vehicles are more prone to rollover accidents with a high center of gravity (CG) and narrow track width such as large vans, pickup trucks, and sport utility vehicles. Therefore, rollover prevention is important for vehicle dynamics and active safety [2].

1.3 FACTORS AFFECTING VEHICLE ROLLOVER

Vehicle rollover accidents may be grouped into two categories: tripped and untripped rollovers. A tripped rollover commonly occurs when a vehicle skids and digs its tires into soft soil or hits a tripping mechanism such as a curb with a sufficiently large lateral velocity. On the other hand, the untripped rollover is induced by extreme maneuvers during critical driving situations, such as excessive speed during cornering, obstacle avoidance, and severe lane change maneuver. In these situations, the forces at the tire-road contact point are enough to cause the vehicle to roll over. Furthermore, vehicle rollover can occur during external disturbances such as side-wind and steering excitation. There is a good body of work in untripped rollovers while studies in tripped rollovers are relatively limited although crash data show that the majority of rollover accidents are tripped rollovers [3]. The reason is that tripped rollovers are more complex, and their dynamics are more complicated and less understood.

1.4   SUMMARY

This book is organized as follows. In Chapter 2, different vehicle dynamic models of rollover are presented, including roll plane model, yaw-roll model, lateral-yaw-roll model, and yaw-roll-vertical model. After that, some precise indexes are proposed and examined to detect vehicle rollover risk under both untripped and tripped situations in Chapter 3 and 4. Then, rollover avoidance control such as anti roll bar, active suspension, active steering system, and differential braking system are studied by researchers in Chapter 5. Chapter 6 gives some control methods used. Finally, some concluding remarks are given.

CHAPTER 2

Dynamic Model of Vehicle Rollover

A proper vehicle model is required to study vehicle rollover dynamics. In recent years, to evaluate the possibility of a vehicle rollover, much efforts have been put on rollover stability. To sum up, several dynamics models used in literature are reviewed: (1) roll plane model, (2) yaw-roll model, (3) lateral-yaw-roll model, (4) yaw-roll-vertical model, (5) multi-freedom model, and (6) multi-body dynamic model.

2.1 ROLL PLANE MODEL

Roll motion is one of the most important causes for vehicle rollover. Therefore, a model describing the roll motion properly is the basis to study vehicle rollover. Many researchers have analyzed vehicle rollover according to the roll plan model [4]. Yu introduced a suspended roll plane model to determine the threat of impending rollover [5]. The proposed roll plane model is shown in Figure 2.1.

Equilibrium for the sprung mass can be derived as follows:

$$\sum F_y = F_{y,RC} \cos \phi_B + F_{z,RC} \sin \phi_B - m_s a_y = 0 \tag{2.1}$$

$$\sum F_z = -F_{y,RC} \sin \phi_B + F_{z,RC} \cos \phi_B - m_s g = 0 \tag{2.2}$$

$$\sum M_{RC} = m_s a_y d \cos \phi_s + m_s g d \sin \phi_s - k_\phi \phi. \tag{2.3}$$

According to the geometrical relationships shown in Figure 2.1,

$$d = (H_{CG,s} - H_{RC}) / \cos \phi. \tag{2.4}$$

Assuming small angles, Equations (2.3) and (2.4) become

$$k_\phi \phi \approx m_s a_y d + m_s g d (\phi - \phi_B) \tag{2.5}$$

and

$$d \approx H_{CG,s} - H_{RC}, \tag{2.6}$$

where a_y is the lateral acceleration of vehicle; d is the distance from CG of sprung mass to roll axis; F_y is the lateral tire force in tire y-axis (of wheel plane); F_z is the vertical force on tire;

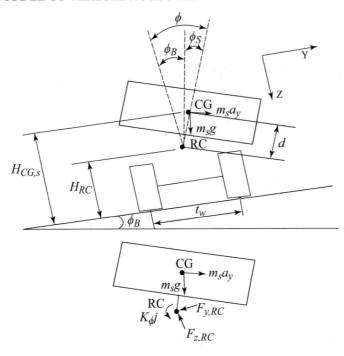

Figure 2.1: Schematic representation of suspended roll plane model.

g is the acceleration of gravity; $H_{CG,s}$ is the CG height of sprung mass respect to grand; H_{RC} is the height of roll axis respect to grand; K_ϕ is the roll stiffness of suspension; M_{RC} is the roll moment, measured about roll center; m_s is the sprung mass; ϕ is the total roll angle of vehicle; and ϕ_B is the road bank angle.

By the roll plane model, a threshold used to determine the degree of rollover can be derived. This model can also be used to design the rollover control strategy.

2.2   YAW-ROLL MODEL

There are many factors that cause the vehicle to roll over. Since it only considers the influence of roll motion on the rollover, the roll model has some limitations. Some researchers consider the yaw motion on the basis of the roll plane model. Yu established a yaw-roll model for heavy-duty vehicle and designed a prototype active roll control system [6]. Chen also used simplified yaw-roll model to compute a Time-To-Rollover index and then corrected it using an Artificial Neural Network [7]. As can be seen from Figure 2.2, the proposed yaw-roll model is separated into a yaw and a roll part. This serial arrangement may have less accurate results compared with an integrated yaw-roll model. However, this simplified structure was found to be superior in two aspects: (1) ease of model construction and (2) faster computations.

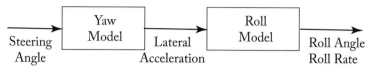

Figure 2.2: Structure of the simplified yaw-roll model.

As shown in Figure 2.3, the vehicle yaw model was assumed to be described by a linear bicycle model and the vehicle speed is assumed to be constant.

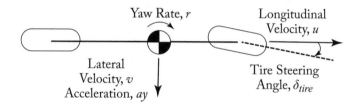

Figure 2.3: Yaw model.

For a linear bicycle model, the discrete-time transfer function from the steering angle to lateral acceleration is known to have the following form:

$$T_{yaw}(z) = \frac{b_0 z^2 + b_1 z + b_2}{z^2 + a_1 z + a_2} = \frac{a_y}{\delta}, \tag{2.7}$$

where a_y is the lateral acceleration and δ is the steering wheel angle which is related to the tire steering angle by a steering gear ratio. After factorization, we can get the following form:

$$T_{yaw}(z) = b_0 + k \frac{z - z_1}{(z - p_1)(z - \bar{p}_1)}, \tag{2.8}$$

where k, z_1, p_1, and p_2 can be calculated by standard system identification techniques and then take the average of their values for multiple files of the same maneuver. These transfer functions have been found to work well under constant speed cases.

The roll model was found to be well behaved which is a 2-degree-of-freedom (DOF) model (sprung mass roll and unsprung mass roll, refer to Figure 2.4).

The structure of the discrete-time transfer fiction from lateral acceleration to sprung mass roll angle is shown as follows:

$$T_{roll}(z) = \frac{b_0 z^3 + b_1 z^2 + b_2 z + b_3}{z^4 + a_1 z^3 + a_2 z^2 + a_3 z + a_4} = \frac{\phi}{a_y}, \tag{2.9}$$

where ϕ is the roll angle of the sprung mass.

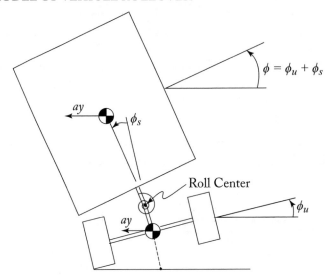

Figure 2.4: Roll model.

Alternatively, Equation (2.9) can be shown in the following form:

$$T_{roll}(z) = k \frac{(z - z_1)(z - z_2)(z - \bar{z}_2)}{(z - p_1)(z - \bar{p}_1)(z - p_2)(z - \bar{p}_2)}, \tag{2.10}$$

where

$$k = \frac{(1 - p_1)(1 - \bar{p}_1)(1 - p_2)(1 - \bar{p}_2)}{(1 - z_1)(1 - z_2)(1 - \bar{z}_2)},$$

z_1, z_2, and $\bar{z}_2$ are zeros, and p_1, $\bar{p}_1$, p_2, $\bar{p}_2$ are pole of $T_{roll}(z)$.

2.3   LATERAL-YAW-ROLL MODEL

Whether the roll plane model or the yaw-roll model can only partially analyze the rollover problem. Therefore, a mathematical model that can fully describe rollover problem is needed. In recent years, the 3-DOF vehicle model including lateral, yaw, and roll motion becomes the most commonly used model [8–12]. The three degrees of freedom vehicle model can be established as follows [13].

For the vehicle model shown in Figure 2.5, a frame of coordinates is fixed on the vehicle body. In studying the rollover dynamics of the vehicle moving at a constant steering angle and a constant speed, the dynamic equations of lateral, yaw, and roll motions can be established as follows:

$$\begin{cases} ma_y - m_s h \ddot{\phi} = 2F_f \cos \delta + 2F_r \\ I_z \dot{r} = 2a F_f \cos \delta - 2b F_r \\ I_x \ddot{\phi} - m_s h a_y = m_s g h \sin \phi - c_\phi \dot{\phi} - k_\phi \phi, \end{cases} \tag{2.11}$$

where the lateral acceleration is given below:

$$a_y = \dot{v} + ur. \tag{2.12}$$

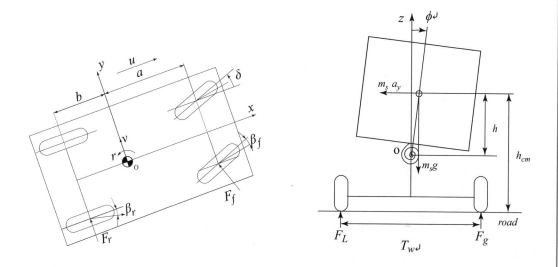

Figure 2.5: Lateral-yaw-roll model.

Equation (2.11) describes the balance relations of the lateral forces on the entire vehicle, the yaw moments of the entire vehicle, and the roll moments on the sprung mass, respectively.

The lateral forces in Equation (2.11) mainly come from the contact between the tire and the road surface at each front and rear wheel and is a function of the physical properties of the tire and the corresponding sideslip angles β_f or β_r observed on the front wheel or rear wheel, respectively. The slip angle of a tire can be determined from the simple geometric relations shown in Figure 2.6 as follows:

$$\beta_f = \arctan\left(\frac{v + ar}{U}\right) - \delta, \qquad \beta_r = \arctan\left(\frac{v - br}{U}\right). \tag{2.13}$$

In this study, a simple tire model with linear constant cornering stiffness will be used so that the lateral forces of tires yield

$$F_f = -k_f \beta_f, \qquad F_r = -k_r \beta_r. \tag{2.14}$$

As the vehicle is moving in cornering, the lateral velocity and yaw rate do not vanish. So, the dynamics of vehicle rollover can be described by Equations (2.11), (2.12), (2.13), and (2.14) in partial unknown state variables v, r, ϕ, and $\dot{\phi}$. That is, the dynamic equation of vehicle

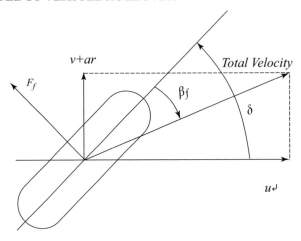

Figure 2.6: Zoom view of a front wheel.

rollover can be recast as

$$\dot{X} = f(X, \gamma), \qquad X, f \in \Re^4, \tag{2.15}$$

where

$$X = (x_1, x_2, x_3, x_4)^T = (v, r, \phi, \dot{\phi})^T, \qquad f \equiv (f_1, f_2, f_3, f_4)^T, \tag{2.16}$$

where

$$f_1 = \frac{\left(-2k_f\beta_f\cos\delta - 2k_r\beta_r\right)I_x + m_s^2 h^2 g\sin\phi - m_s h k_\phi\phi - m_s h c_\phi\dot{\phi} - \left(mI_x - m_s^2 h^2\right)Ur}{mI_x - m_s^2 h^2}$$

$$f_2 = \frac{2bk_r\beta_r - 2ak_f\beta_f\cos\delta}{I_z}, \qquad f_3 = x_4$$

$$f_4 = \frac{\left(-2k_f\beta_f\cos\delta - 2k_r\beta_r\right)m_s h + m_s h m g\sin\phi - m k_\phi\phi - m c_\phi\dot{\phi}}{mI_x - m_s^2 h^2}$$

$$\gamma \equiv \left(\delta, U, m, m_s, h, I_x, I_z, k_f, k_r, k_\phi, c_\phi, a, b\right)^T.$$

In these equations, a and b represents longitudinal distance from the center of gravity to the front and the rear axle; c_φ is equivalent roll damping coefficient of suspension; F_f/F_r is lateral force of a front/rear tire; F_L/F_R is vertical load on left/right tires; h is the length of roll arm measured from the center of gravity to the roll center; I_x represents roll moment of inertia of the sprung mass, measured about the roll axis; I_z represents yaw moment of inertia of the total mass, measured about the z axis; k_f/k_r represents cornering stiffness coefficient of a front/rear tire; k_ϕ is the equivalent roll stiffness coefficient of suspension; L is the wheelbase of vehicle; m is the total mass of vehicle; m_s is sprung mass of vehicle; r is yaw rate of the sprung mass; T_w is track width of vehicle; U is forward speed of vehicle; v is lateral speed of vehicle; δ represents steering angle of front wheels; ϕ is the roll angle; $\dot{\phi}$ is the roll rate; and $\ddot{\phi}$ is roll acceleration.

2.4   YAW-ROLL-VERTICAL MODEL

The above-mentioned 3-DOF model only applies to untripped rollover. However, most of the rollover accidents in real life are tripped rollover, and at present, there is little research on tripped rollover. So, some researchers proposed yaw-roll-vertical model which takes vertical road excitation [14]. A new rollover index was proposed based on this model to evaluate the possibility of vehicle rollover under both untripped and special tripped situations. Jin et al. established vehicle model which consists of the lateral and yaw motions, and the roll and vertical motions of sprung and two unsprung masses [15], as shown in Figure 2.7.

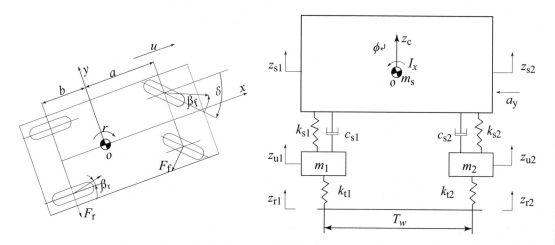

Figure 2.7: 6-DOF model of vehicle.

In order to simplify the rollover stability model, some assumptions are made as follows. The pitch dynamics can be ignored, and the front and the rear steering angles are small. Also, the properties of tire are regarded as symmetric with respect to x-axis. Furthermore, the effects of the lateral wind, the longitudinal motion, and the roll motion of unsprung mass are neglected since they are of the secondary importance.

From D'Alembert's principle, the equations of the above model are as follows.
Lateral motion:

$$ma_y - m_s h\ddot{\phi} = 2F_f + 2F_r. \tag{2.17}$$

Yaw motion:

$$I_z \dot{r} = 2a F_f - 2b F_r + M_B. \tag{2.18}$$

Roll motion:

$$I_x \ddot{\phi} = (F_{s2} - F_{s1}) \frac{T_w}{2} + m_s h a_y + m_s h g \phi. \tag{2.19}$$

Vertical motion of sprung mass:

$$m_s \ddot{z}_c = F_{s1} + F_{s2}. \tag{2.20}$$

Vertical motions of unsprung masses:

$$m_1 \ddot{z}_{u1} = -F_{s1} - k_{t1}(z_{u1} - z_{r1}) \tag{2.21}$$

$$m_2 \ddot{z}_{u2} = -F_{s2} - k_{t2}(z_{u2} - z_{r2}). \tag{2.22}$$

In these equations, m_1 and m_2 are the left unsprung mass and the right unsprung mass, respectively; k_{t1} and k_{t2} are the vertical stiffness of the left tires and right tires, respectively; z_c is the vertical displacement of sprung mass; z_{u1} and z_{u2} are the vertical displacement of the left unsprung mass and the right unsprung mass, respectively; z_{r1} and z_{r2} are the road input of the left tires and the right tires, respectively; F_{s1} and F_{s2} are the dynamic forces of the left suspension and the right suspension due to vertical acceleration, respectively; and M_B is the anti-yaw torque.

The dynamic forces of the left and right suspensions due to vertical acceleration can be written as:

$$F_{s1} = -k_{s1}(z_{s1} - z_{u1}) - c_{s1}(\dot{z}_{s1} - \dot{z}_{u1}) \tag{2.23}$$

$$F_{s2} = -k_{s2}(z_{s2} - z_{u2}) - c_{s2}(\dot{z}_{s2} - \dot{z}_{u2}). \tag{2.24}$$

In Equations (2.23) and (2.24), k_{s1} and k_{s2} are the vertical stiffness of the left suspension and the right suspension, respectively; c_{s1} and c_{s2} are the equivalent damping coefficient of the left suspension and the right suspension; and z_{s1} and z_{s2} are the vertical displacement of sprung mass on the left and on the right, respectively.

By taking the coupling relationship between the vertical motion and lateral motion of the sprung mass into consideration, the equation can be obtained as follows:

$$\begin{bmatrix} z_{s1} \\ z_{s2} \end{bmatrix} = G^T \begin{bmatrix} z_c \\ \varphi \end{bmatrix}, \tag{2.25}$$

where

$$G = \begin{bmatrix} 1 & 1 \\ -T_w/2 & T_w/2 \end{bmatrix}.$$

2.5   MULTI-FREEDOM MODEL

Given the exclusive features of heavy-duty vehicles such as the high center of gravity, the big wheel tread, the long wheelbase, the large number of passengers' capacity, and the variable distribution of passengers having an impact on its rollover property, the above-mentioned rollover model cannot accurately describe the roll motion. Therefore, it is necessary to establish a multi-freedom rollover dynamics model to represent the motion state of heavy-duty vehicles. In this section, a six degree of freedom rollover dynamics model is established for a triaxle bus which has complex structure. For a triaxle bus, the middle axle and the rear axle are on the same side of the center of mass, and the distance between the middle axle and the rear axle is short such that the roll coupling between the middle axle and the rear axle is neglected. Therefore, the middle axle and the rear axle of the triaxle bus is equivalent to a virtual rear axle, as shown in Figure 2.8.

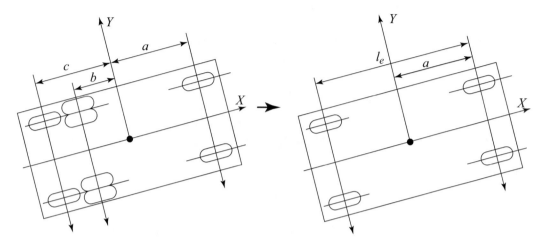

Figure 2.8: The equivalent model of the triaxle bus.

In addition, a twisted bar with a constant stiffness is assumed to link between the first axle and the virtual rear axle. For the sake of simplicity, the effects of the lateral wind, the pitching motion, and the longitudinal motion are neglected since they are of secondary importance in studying the rollover of such a vehicle, the road profile is regarded as symmetric with respect to the x axle. Thus, a 6-DOF vehicle model moving at a constant speed and constant steering angle is established, as shown in Figure 2.9.

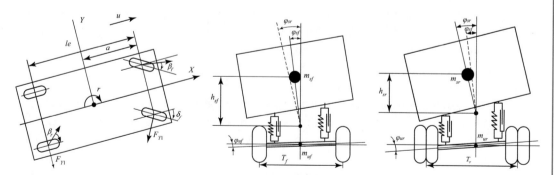

Figure 2.9: Dynamic model of the triaxle bus rollover.

From D'Alembert's principle, the equations of the above model are as follows. Lateral motion:

$$ma_y - m_{sf}h_f\ddot{\varphi}_{sf} - m_{sr}h_r\ddot{\varphi}_{sr} = 2F_f\cos\delta + 2F_{Yr}. \qquad (2.26)$$

Yaw motion:

$$I_Z \dot{r} = 2a F_f \cos \delta + M_r.$$
(2.27)

Roll motion of the sprung mass of the front axle:

$$
\begin{aligned}
I_{Xf} \ddot{\varphi}_{sf} = {} & m_{sf} h_f a_y + m_{sf} g h_f \varphi_{sf} - k_f \left(\varphi_{sf} - \varphi_{uf} \right) \\
& - l_f \left(\dot{\varphi}_{sf} - \dot{\varphi}_{uf} \right) + k_b \left(\varphi_{sf} - \varphi_{sr} \right).
\end{aligned}
$$
(2.28)

Roll motion of the sprung mass of the rear axle:

$$
\begin{aligned}
I_{Xr} \ddot{\varphi}_{sr} = {} & m_{sr} h_r a_y + m_{sr} g h_r \varphi_{sr} - k_r \left(\varphi_{sr} - \varphi_{ur} \right) \\
& - l_r \left(\dot{\varphi}_{sr} - \dot{\varphi}_{ur} \right) - k_b \left(\varphi_{sr} - \varphi_{sf} \right).
\end{aligned}
$$
(2.29)

Roll motion of the unsprung mass of the front axle:

$$
\begin{aligned}
2F_{Y1} h_c + m_{uf} \left(h_{uf} - h_{cf} \right) a_y = {} & k_{uf} \varphi_{uf} - m_{uf} g \left(h_{uf} - h_{cf} \right) \varphi_{uf} \\
& - k_f \left(\varphi_{sf} - \varphi_{uf} \right) - l_f \left(\dot{\varphi}_{sf} - \dot{\varphi}_{uf} \right).
\end{aligned}
$$
(2.30)

Roll motion of the unsprung mass of the virtual rear axle:

$$
\begin{aligned}
2 \left(F_{Y2} + F_{Y3} \right) h_c + m_{ur} \left(h_{ur} - h_{cr} \right) a_y = {} & k_{ur} \varphi_{ur} - m_{ur} g \left(h_{ur} - h_{cr} \right) \varphi_{ur} \\
& - k_r \left(\varphi_{sr} - \varphi_{ur} \right) - l_r \left(\dot{\varphi}_{sr} - \dot{\varphi}_{ur} \right),
\end{aligned}
$$
(2.31)

where

$$
\begin{cases}
F_{Yr} = F_m + F_r \\
M_r = 2b_1 F_m + 2c_1 F_r.
\end{cases}
$$
(2.32)

In the above-mentioned equations, m_{sf} represents the equivalent sprung mass of the front axle; m_{sr} indicates the equivalent sprung mass of the rear axle; m_{uf} refers to the unsprung mass of the front axle; m_{ur} is the unsprung mass of the rear two axles; b_1 and c are the longitudinal distance from the CG to the middle axle and rear axle, respectively; h_f is the height between the center of front sprung mass and the roll center; h_r is the height between the center of rear sprung mass and the roll center; h_{uf} and h_{ur} are the height of the center of the front unsprung mass and the rear unsprung mass, measured upward from the road; h_{cf} and h_{cr} are the height of the front roll center and the rear roll center, measured upward from the road, respectively; I_{Xf} and I_{Xr} are the roll inertia of the front sprung mass and the rear sprung mass, measured about the roll axle; φ_{sf} and φ_{sr} are the roll angle of the front sprung mass and the rear sprung mass; φ_{uf} and φ_{ur} are the roll angle of the front unsprung mass and the rear unsprung mass; F_{Yr} the lateral force of the tires at the virtual axle; M_r is the yaw moment caused by the virtual rear axle; F_f, F_m, and F_r are the lateral force of the tires at the first axle, the middle axle, and the rear axle, respectively; k_f and k_r are the equivalent roll stiffness coefficient of the front suspension and the rear suspension; k_{uf} and k_{ur} are the equivalent roll stiffness coefficient of the front unsprung mass and the rear unsprung mass; l_f and l_r are the equivalent roll damping coefficient of the front suspension and the rear suspension; and k_b is the torsion stiffness coefficient of vehicle frame.

In addition, the steering angle of front wheels δ is assumed to be sufficiently small that $\cos\delta \approx 1$ in Equations (2.26) and (2.27) holds.

The lateral forces in Equations (2.26) and (2.27) mainly come from the contact between the road and tires at the front, middle, and rear axle, depending on the physical properties of the tire and the corresponding side slip angles β_f, β_m, and β_r observed on the front wheels, middle wheels, and rear wheels, respectively. In addition, the two wheels at the front axle will rotate around the king bolt, the two middle wheels and rear wheels rotate around the axle vertical to the road, due to the roll motion. Therefore, the slip angle of a tire can be determined from the simple geometric relations, as follows:

$$
\begin{cases}
\beta_f = \arctan\left(\dfrac{v + ar}{U}\right) - \delta \\[2mm]
\beta_m = \arctan\left(\dfrac{v - br}{U}\right) \\[2mm]
\beta_r = \arctan\left(\dfrac{v - cr}{U}\right).
\end{cases}
\tag{2.33}
$$

Using a simple tire model with a linear constant cornering stiffness, the lateral forces of tires can be obtained:

$$
\begin{cases}
F_f = -k_f \beta_f \\
F_m = -k_m \beta_m \\
F_r = -k_r \beta_r,
\end{cases}
\tag{2.34}
$$

where k_f, k_m, and k_r are the cornering stiffness of the front wheels, the middle wheels, and the rear wheels, respectively.

Setting $U_s = \begin{bmatrix} \varphi_{sf} & \varphi_{sr} & \varphi_{uf} & \varphi_{ur} \end{bmatrix}^T$, $V = \begin{bmatrix} \dot{\varphi}_{sf} & \dot{\varphi}_{sr} \end{bmatrix}^T$, and substituting the Equations (2.33) and (2.34) into Equations (2.26)–(2.32), the state space equations of vehicle rollover system can be obtained in a matrix form as follows:

$$
M_q \begin{bmatrix} \dot{v} \\ \dot{r} \\ \dot{U}_s \\ \dot{V} \end{bmatrix} = A_q \begin{bmatrix} v \\ r \\ U_s \\ V \end{bmatrix} + B_q \delta,
\tag{2.35}
$$

where

$$M_q = \begin{bmatrix} M_1 & 0_{2\times 4} & M_2 \\ M_3 & M_4 & M_5 \\ 0_{2\times 2} & M_6 & 0_{2\times 2} \end{bmatrix}; \qquad A_q = \begin{bmatrix} A_1 & 0_{2\times 4} & 0_{2\times 2} \\ A_2 & A_3 & A_4 \\ 0_{2\times 2} & 0_{2\times 4} & A_5 \end{bmatrix};$$

$$B_q = \begin{bmatrix} 2K_f & 2aK_f & 0 & 0 & 2h_f K_f & 0 & 0 & 0 \end{bmatrix}^T;$$

$$M_1 = \begin{bmatrix} m & 0 \\ 0 & I_Z \end{bmatrix}; \qquad M_2 = \begin{bmatrix} -h_f m_{sf} & -h_r m_{sr} \\ 0 & 0 \end{bmatrix};$$

$$M_3 = \begin{bmatrix} h_f m_{sf} & 0 \\ h_r m_{sr} & 0 \\ -m_{uf}(h_{uf} - h_{cf}) & 0 \\ -m_{ur}(h_{ur} - h_{cr}) & 0 \end{bmatrix}; \qquad M_4 = \begin{bmatrix} 0 & 0 & 0 & 0 \\ 0 & 0 & 0 & 0 \\ 0 & 0 & l_f & 0 \\ 0 & 0 & 0 & l_r \end{bmatrix};$$

$$M_5 = \begin{bmatrix} -I_{Xf} & 0 \\ 0 & -I_{Xr} \\ 0 & 0 \\ 0 & 0 \end{bmatrix}; \qquad M_6 = \begin{bmatrix} 1 & 0 & 0 & 0 \\ 0 & 1 & 0 & 0 \end{bmatrix};$$

$$A_1 = \begin{bmatrix} \dfrac{-(2K_f + 2K_m + 2K_r)}{U} & \dfrac{-(mu^2 + 2aK_f - 2b_1 K_m - 2c_1 K_r)}{U} \\ \dfrac{-(2aK_f - 2bK_m - 2cK_r)}{U} & \dfrac{-(2a^2 K_f + 2b_1^2 K_m + 2c_1^2 K_r)}{U} \end{bmatrix};$$

$$A_2 = \begin{bmatrix} 0 & -h_f m_s U \\ 0 & -h_r m_{sr} U \\ \dfrac{-2K_f h_c}{U} & \dfrac{m_{uf} U^2 (h_{uf} - h_{cf}) - 2aK_f h_{cf}}{U} \\ \dfrac{-2K_m + K_r)h_c}{U} & \dfrac{m_{ur} U^2 (h_{ur} - h_{cr}) + 2bK_m h_{cr} + 2cK_r h_{cr}}{u} \end{bmatrix};$$

$$A_3 = \begin{bmatrix} \begin{matrix} k_b + k_f \\ -m_{sf}gh_f \end{matrix} & -k_b & -k_f & 0 \\ -k_b & \begin{matrix} k_b + k_r \\ -m_{sr}gh_r \end{matrix} & 0 & -k_r \\ k_f & 0 & \begin{matrix} -k_{uf} + m_{uf}g \\ (h_{uf} - h_{cf}) - k_f \end{matrix} & 0 \\ 0 & k_r & 0 & \begin{matrix} -k_{ur} + m_{ur}g \\ (h_{ur} - h_{cr}) - k_r \end{matrix} \end{bmatrix};$$

$$A_4 = \begin{bmatrix} l_f & 0 \\ 0 & l_r \\ l_f & 0 \\ 0 & l_r \end{bmatrix}; \qquad A_5 = \begin{bmatrix} 1 & 0 \\ 0 & 1 \end{bmatrix}.$$

Because the vehicle moves when cornering, the lateral velocity and yaw rate do not vanish. Hence, the dynamics of vehicle rollover can be described by Equation (2.34) in the partial unknown state variables v, r, and vector U_s, V. Setting the state vector as $x = \begin{bmatrix} \dot{v} & \dot{r} & \dot{U}_s & \dot{V} \end{bmatrix}^T$. Then Equation (2.35) can be rewritten into (2.36):

$$\dot{x} = Ax + B\delta, \tag{2.36}$$

where

$$A = M_q^{-1} \times A_q, \qquad B = M_q^{-1} \times B_q.$$

The distance between the first axle and the virtual rear axle is the equivalent wheelbase. According to some papers, the method to determine the equivalent wheelbase is obtained. First, the linear 2-DOF model of the vehicle can be set up based on Equations (2.25) and (2.26),

$$\begin{cases} \begin{bmatrix} \dot{\beta} \\ \dot{r} \end{bmatrix} = \begin{bmatrix} \dfrac{-k_f - k_m - k_r}{mU} & \dfrac{-ak_f + b_1k_m + c_1k_r}{mU^2} - 1 \\ \dfrac{-ak_f + b_1k_m + c_1k_r}{I_Z} & \dfrac{-a^2k_f - b_1^2k_m - c_1^2k_r}{I_Z U} \end{bmatrix} \begin{bmatrix} \beta \\ r \end{bmatrix} + \begin{bmatrix} \dfrac{k_f}{mU} \\ \dfrac{ak_f}{I_Z} \end{bmatrix} [\delta] \\ \beta = \dfrac{v}{U}. \end{cases}$$
$$\tag{2.37}$$

The yaw rate is fixed at steady state, now $\dot{\beta} = 0$ and $\dot{r} = 0$, so

$$\begin{bmatrix} \beta \\ r \end{bmatrix} = \begin{bmatrix} \dfrac{-k_f - k_m - K_r}{mU} & \dfrac{-ak_f + b_1k_m + c_1k_r}{mU^2} - 1 \\ \dfrac{-ak_f + b_1k_m + c_1k_r}{I_Z} & \dfrac{-a^2k_f - b_1^2k_m - c_1^2k_r}{I_Z U} \end{bmatrix}^{-1} \begin{bmatrix} \dfrac{k_f}{mU} \\ \dfrac{ak_f}{I_Z} \end{bmatrix} [\delta]. \tag{2.38}$$

When only taking the front axle steering into account, the yaw rate gain can be described as follows:

$$\frac{r}{\delta} = \frac{u\left[k_f k_m (a + b_1) - k_f k_r (a + c_1)\right]}{\left\{\begin{array}{l} k_f k_m (a + b_1)^2 + k_f k_r (a + c_1)^2 + k_m k_r (c_1 - b_1)^2 \\ -mU^2 (ak_f - b_1 k_m - c_1 k_r) \end{array}\right\}}. \tag{2.39}$$

Setting

$$\begin{cases} l = a + c_1 \\ t = c_1 - b_1. \end{cases} \tag{2.40}$$

The yaw rate gain can be simplified as follows:

$$\frac{r}{\delta} = \frac{U}{\left\{\begin{array}{l} \left[k_f k_m (l - t)^2 + k_f k_r l^2 + k_m k_r t^2\right] / \left[k_f k_m (l - t) + k_f k_r l\right] \\ +U^2 \left[-m (ak_f - b_1 k_m - c_1 k_r)\right] / \left[k_f k_m (l - t) + k_f k_r l\right] \end{array}\right\}}. \tag{2.41}$$

So, the equivalent wheelbase of the triaxle bus can be obtained according to the equivalence of physical meaning as follows:

$$l_e = \frac{k_f k_m (l - t)^2 + k_f k_r l^2 + k_m k_r t^2}{\left[k_f k_m (l - t) + k_f k_r l\right]}. \tag{2.42}$$

Then, the equivalent front sprung mass is

$$m_{sf} = \frac{m_s (l_e - a)}{l_e} \tag{2.43}$$

and the equivalent rear sprung mass is

$$m_{sr} = \frac{m_s a}{l_e}. \tag{2.44}$$

Also, the equivalent front unsprung mass and the equivalent rear unsprung mass can be described as

$$\begin{cases} m_{uf} = \dfrac{m_u (l_e - a)}{l_e} \\ m_{ur} = \dfrac{m_s a}{l_e}. \end{cases} \tag{2.45}$$

2.6   MULTI-BODY DYNAMIC MODEL

Experimental testing to improve safety is accurate but it is also expensive and dangerous. Therefore, multi-body dynamic model is used by researchers to improve the understanding of rollover dynamics [16–20]. Typically, Pawel used a complex and highly nonlinear multi-body model

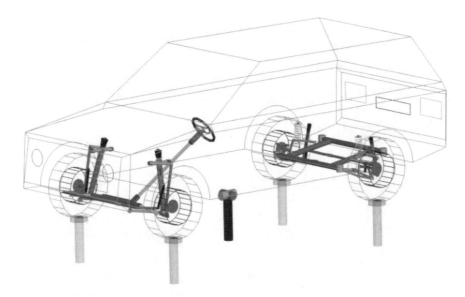

Figure 2.10: Multi-body dynamics model.

with 165 DOF which is correlated to vehicle kinematic and compliance (K&C) measurements. Furthermore, the Magic Formula tyre model is employed, as shown in Figure 2.10.

This model used 6 DOFs to represent vehicle body, 78 DOFs to represent front axle suspension and 76 DOFs to represent rear axle suspension, 1 DOF to represent steering system, and 4 DOFs to represent wheels.

In recent years, the multi-body dynamics can be obtained by some business software such as Trucksim, Carsim, and Adams. According to these softwares, the cost and risk are greatly decrease. At the same time, the influence of vehicle driving parameters, and structure parameters can also be studied.

Bao and Hu [21] built the multi-body dynamics model based on Adams to analyze the rollover stability considering road excitation, as shown in Figure 2.11. And the simulation analysis of rollover stability was carried out. Figures 2.12 and 2.13 show the model obtained by Trucksim and Carsim, which aim at heavy and light vehicles, respectively.

2.7 SUMMARY

Until now, many researchers have studied the vehicle rollover dynamic models. In addition to the model above, there are many other vehicle rollover dynamic models established in order to more accurately describe the vehicle rollover motion state. However, little research has been done on the modeling of tripped rollover.

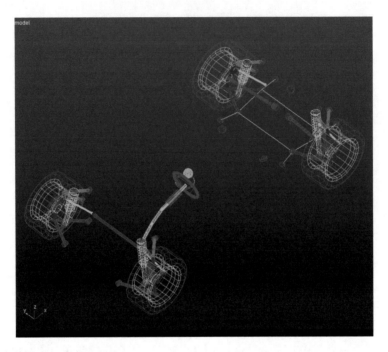

Figure 2.11: Adams model.

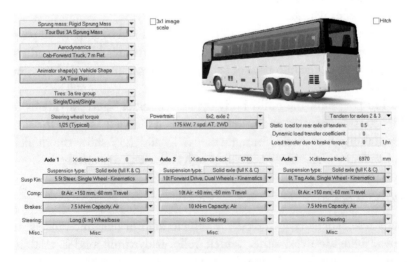

Figure 2.12: Trucksim model.

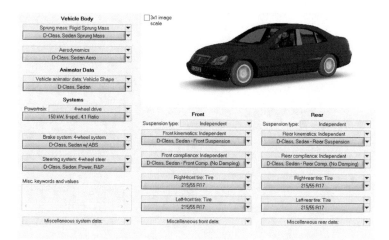

Figure 2.13: **Carsim model.**

CHAPTER 3

Stability of Untripped Vehicle Rollover

To help mitigate the occurrence of rollover accidents at problem locations, rollover detection and warning is necessary. At a minimum, problem locations may have speed reduction signs, rollover warning signs, and chevrons to pre-warn all drivers of the potentially dangerous curve ahead. Although signing is present at most problem locations, many drivers become desensitized to a specific warning amongst the multitude of other roadside signs [22]. So, accurate and real-time detection and warning of the impending danger of a vehicle rollover is important for improving vehicle active safety and reduce the occurrence of rollover accidents. In this chapter, the roll indexes and warning for untripped vehicle rollover is studied and analyzed.

3.1 ROLL INDEX OF UNTRIPPED VEHICLE ROLLOVER

This section introduces some typical rollover evaluation method which can predict untripped vehicle rollover includes static stability factor, dynamics stability factor, and lateral load transfer ratio.

3.1.1 STATIC STABILITY FACTOR

Many passive warning systems use a prediction algorithm to determine the threat of impending rollover based on the values of vehicle roll angle and/or lateral acceleration. In order to help measure the likelihood of vehicle rollover, a rollover resistance rating program was proposed by NHTSA, which uses the Static Stability Factor (SSF) [4, 5, 23]. In SSF, the stiffness and damping of suspension system and tires are ignored. Applying torque balance equation to one side tires of the vehicle, the following equation can be derived:

$$ma_y H - mg \sin \phi_B H + F_{zi} T_w - \frac{1}{2} mg T_w, \qquad (3.1)$$

where ϕ_B is road ramp angle, H represents the height of center of mass, measures upward from the road, F_{zi} is vertical force of the other side tires, and T_w is the track width.

Assuming the road ramp angle is zero, when the vertical force on the other side tire of vehicle is reduced to 0, being $F_{zi} = 0$, it is believed that the vehicle will roll over. Then the SSF can be obtained from Equation (3.1) as follows:

$$SSF = \frac{a_y}{g} = \frac{T_w}{2H}. \tag{3.2}$$

It can be seen from Equation (3.2) that the value of SSF only relates to track and height of center of mass, which is convenient for application. Ronald Huston [24] used SSF and "dynamic stability factor" which is defined in the same way as the SSF but for the dynamic case to predict rollover. Figures 3.1 and 3.2 show the variation of vehicle body critical title angle θ_{cr} with SSF and with the tire/road friction μ [24]. As expected, the greater the SSF value, the larger is the tilt angle, and thus the less likely the vehicle is to roll over and, conversely, the smaller the SSF, the greater is the rollover propensity.

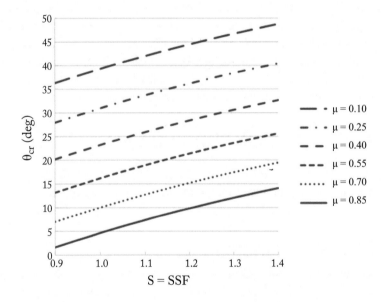

Figure 3.1: Critical tilt angle as a function of the SSF for various values of the coefficient of friction.

This measure of rollover propensity reflects only the most fundamental relation and does not take the effects of suspension and tire compliance into account. It cannot effectively predict rollover risk during a dynamic case. Therefore, it is essential to propose a rollover index which applies to both steady case and dynamic case.

3.1.2   DYNAMICS STABILITY FACTOR

Because the SSF does not work well in a dynamic condition. To detect wheel liftoff conditions when a vehicle is moving, Jin et al. defined a Dynamics Stability Factor (DSF) based on

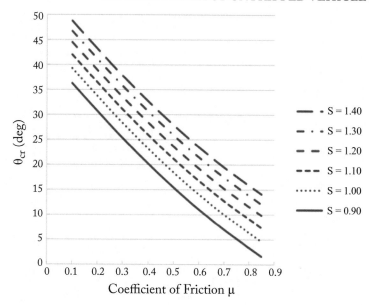

Figure 3.2: Critical tilt angle as a function of the coefficient of friction for various values of the SSF.

SSF [13]:

$$DSF = \frac{T_w}{2H} - \frac{u^2 m_s h^2 \delta}{LH\left(k_\phi - m_s hg\right)\left[1 - \dfrac{mu^2}{L^2}\left(\dfrac{a}{2k_r} - \dfrac{b}{2k_f}\cos\delta\right) - \dfrac{u^2 m_s h\left(c_f - c_r\right)}{L\left(k_\phi - m_s hg\right)}\right]}. \quad (3.3)$$

In the steady state motion, the lateral acceleration at the rollover threshold can be predicted by the value of DSF. The vehicle with a larger DSF implies a higher threshold of rollover. Compared with the traditional SSF, the dynamic stability factor has a number of features as follows.

1. From the conditions in Equation (3.3), the second item of DSF (including minus) is always negative. Hence, DSF is always smaller than SSF, and can predict the trend of vehicle rollover more precisely than SSF, as shown in Figure 3.3.

2. DSF includes the effects of the track width T_w and the height of the center of gravity H on vehicle dynamic stability. DSF increases with an increase in the track width and the decrease in the height of the center of gravity. That is, increasing the track width and decreasing the height of the center of gravity can improve the stability of vehicle rollover. This fact gets an agreement with the traditional SSF. As shown in Equation (3.3), the

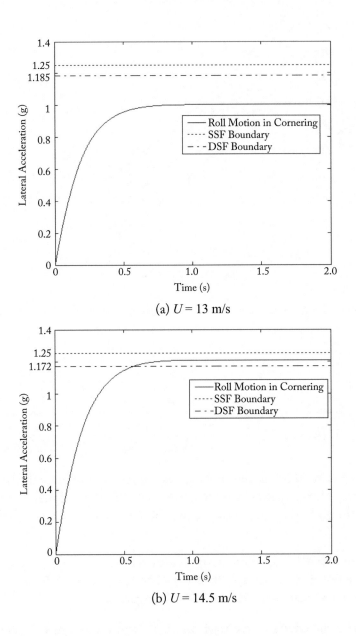

(a) $U = 13$ m/s

(b) $U = 14.5$ m/s

Figure 3.3: Lateral acceleration of roll motion at three forward speeds. (*Continues.*)

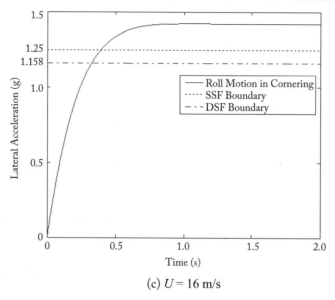

(c) $U = 16$ m/s

Figure 3.3: (*Continued.*) Lateral acceleration of roll motion at three forward speeds.

second item of DSF has nothing to do with the track width T_w, but inversely relates to the center of gravity H. Thus, decreasing H receives better improvement on vehicle stability than increasing T_w, as shown in Figure 3.4, while SSF does not show this tendency.

3. DSF takes the longitudinal location of the center of gravity into consideration. As section (b), when wheelbase L is fixed as constant, substituting $b = L - a$ into the second item of DSF yields a function $f(a)$. This is a decreasing function to variable a. Therefore, moving the center of the sprung mass close to the front axle can improve the stability of vehicle rollover, as shown in Figure 3.5.

4. DSF takes the forward speed and the steering angle into account. From Equation (3.3), DSF increases with the decreasing of the forward speed U and the front wheel-steering angle δ. Therefore, this evidence enables one to improve the stability of vehicle rollover by using either a low speed or a small steering angle, as shown in Figure 3.6.

5. DSF varies from the equivalent roll stiffness of the suspension, as seen from Equation (3.3). It is reasonable to expect that enhancing this fact enables one to improve the stability of vehicle rollover, as shown in Figure 3.7.

6. Furthermore, DSF includes the effects of the properties of tires on vehicle dynamic stability. The ratio of the cornering stiffness of a front tire and the rear as $\mu_1 = k_f / k_r$ and substituting it into the second item of DSF yields a function $f(\mu_1)$, which is a decreas-

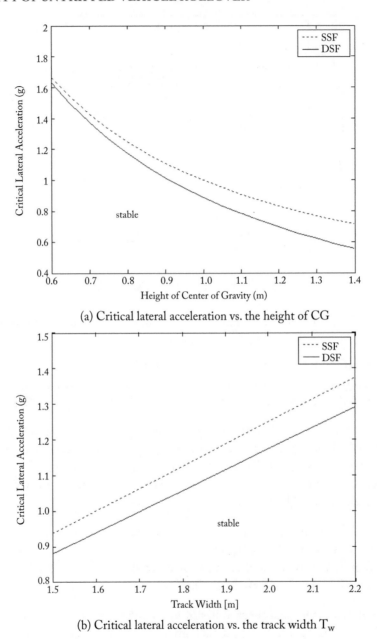

(a) Critical lateral acceleration vs. the height of CG

(b) Critical lateral acceleration vs. the track width T_w

Figure 3.4: The critical lateral acceleration with respect to the height of center of gravity and the track width, respectively, for SSF and DSF.

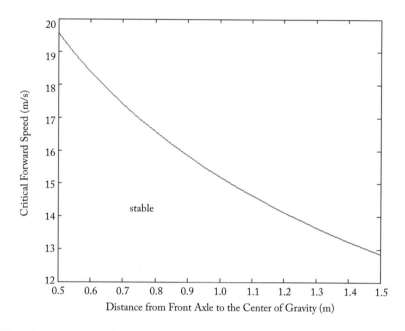

Figure 3.5: Stability region on (a, U_c) plane.

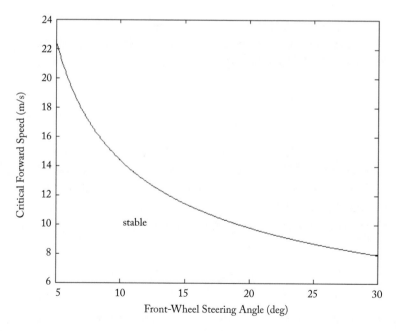

Figure 3.6: Stability region on (δ, U_c) plane.

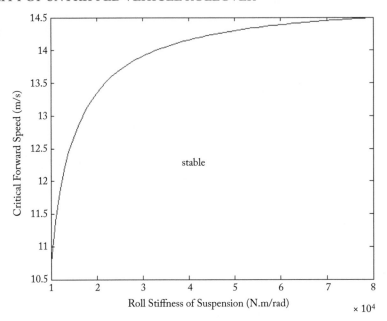

Figure 3.7: Stability region on (k_ϕ, U_c) plane.

ing function to variable μ_1. Hence, DSF increases with the decreasing of μ_1, as shown in Figure 3.8. DSF varies from the steer coefficient c_f and c_r induced by roll, as seen from Equation (3.3). DSF will be increased by decreasing c_f or increasing c_r, as shown in Figures 3.9–3.10.

3.1.3   LATERAL LOAD TRANSFER RATIO

At present, the Lateral Load Transfer Ratio (LTR) is the most widely used evaluation index to predict rollover risk [13, 25–28]. When vehicle is running, the vertical load of the vehicle will gradually shift from the inside to the outside. So, the LTR evaluates the rollover stability according to the vertical load of the wheels during the driving. It is believed that the vehicle will roll over, when the vertical load on one side of the wheel is reduced to 0. The fundamental definition of the LTR is described as follows:

$$LTR = \frac{F_{z1} - F_{z2}}{F_{z1} + F_{z2}},\tag{3.4}$$

where F_{z1} and F_{z2} are the total vertical loads of the left and right wheels of vehicle, respectively. A vehicle is considered to roll over when LTR is more than 1 or less than -1. It should be noted that, F_{z1} equals to F_{z2} and $LTR = 0$ when a vehicle is traveling straight. If $F_{z1} = 0$, then

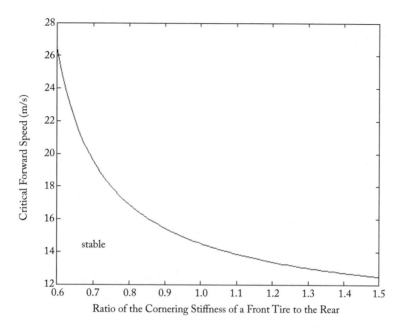

Figure 3.8: Stability region on (μ_1, U_c) plane.

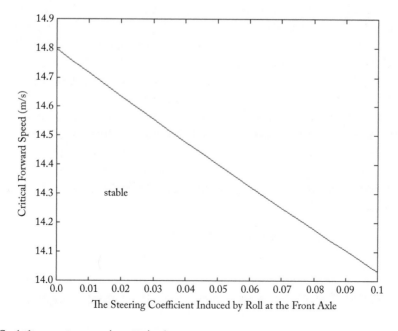

Figure 3.9: Stability region on (c_f, U_c) plane.

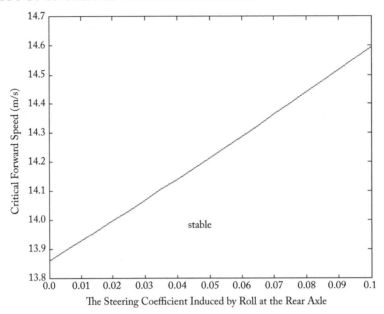

Figure 3.10: Stability region on (c_r, U_c) plane.

$LTR = -1$ and the left wheel just lifts off the ground. Also, if $F_{z2} = 0$, then $LTR = 1$ and the right wheel just lifts off the ground.

Figure 3.11 shows the rollover stability at different conditions (J-turn, Fishhook, and Double Lane Change maneuvers) and different vehicle speed (60 km/h, 80 km/h, and 100 km/h) based on the LTR. It is observed that LTR can effectively predict the vehicle rollover. It also shows that the higher the speed, and the more likely it is to be rollover.

LTR is considered as a very useful index to study the dynamics and simulation of vehicle rollover, while the vertical load of each wheel of the vehicle is difficult to be measured or estimated in real time. So, the LTR cannot be used to predict the risk of vehicle rollover directly, especially in the case of emergency. Many researchers derived new rollover indexes based on the vehicle rollover dynamics model and the definition of the LTR [15, 29–36]. For example, ignore the vertical motion of the vehicle, so

$$\begin{cases} F_{z1} - F_{z2} = F_{s2} - F_{s1} \\ F_{z1} + F_{z2} = mg. \end{cases} \tag{3.5}$$

F_{s1} and F_{s2} are the left and right supporting force of suspension. According to moment equilibrium equations, the difference value between the left and right suspension force can be described as follows:

$$\frac{T_w}{2}(F_{s2} - F_{s1}) = -k_\phi (\phi_s - \phi_u) - c_\phi (\dot{\phi}_s - \dot{\phi}_u), \tag{3.6}$$

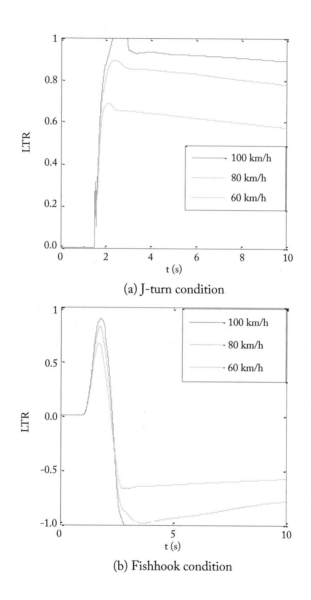

(a) J-turn condition

(b) Fishhook condition

Figure 3.11: Stability analysis at typical conditions. (*Continues.*)

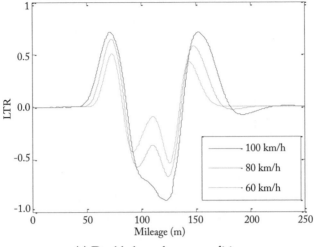

(c) Double lane change condition

Figure 3.11: (*Continued.*) Stability analysis at typical conditions.

where k_ϕ is equivalent roll damping coefficient of suspension, c_ϕ is equivalent roll damping coefficient of suspension, and ϕ_s/ϕ_u represents roll angle of sprung/unsprung mass.

The torque balance of the vehicle under roll motion of sprung mass is given as below:

$$I_x\ddot{\phi}_s = m_s h a_y + m_s h g \phi_s - k_\phi\left(\phi_s - \phi_u\right) - c_\phi\left(\dot{\phi}_s - \dot{\phi}_u\right), \tag{3.7}$$

where I_x is the roll inertia of sprung mass, m_s is sprung mass, h is the height between the center of sprung mass and the roll center, and a_y is the lateral acceleration of the vehicle.

Thus,

$$F_{s2} - F_{s1} = -\frac{2}{T_w}\left[I_x\ddot{\phi}_s - m_s h a_y - m_s h g \phi_s\right]. \tag{3.8}$$

So, the rollover index (RI) can be represented as

$$RI = \frac{-\dfrac{2}{T_w}\left[I_x\ddot{\phi}_s - m_s h a_y - m_s h g \phi_s\right]}{mg}. \tag{3.9}$$

Figure 3.12 shows the comparison between fundamental LTR and new rollover index RI in J-turn and Fishhook condition. In Figure 3.12a for J-turn maneuver, the maximum error between fundamental LTR and RI is about 5%. And the peak value of RI is a little bigger than LTR. That means RI can be more sensitive. In Figure 3.12b, there is a good fit between RI and LTR for Fishhook maneuver. In general, the new rollover index RI agrees with the fundamental definition of LTR when the vehicle rolls with all wheels keeping on road due to

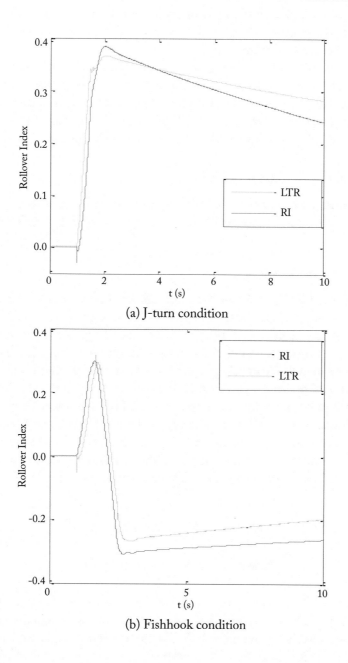

(a) J-turn condition

(b) Fishhook condition

Figure 3.12: Comparison of new rollover index RI and fundamental LTR.

J-turn and Fishhook condition. And at all time, the new rollover index changes more sensitively than fundamental LTR which is obtained using wheel vertical loads information. So, the validity of new rollover index can be demonstrated.

In recent years, with the deepening of the research on vehicle rollover, many researchers have improved the fundamental LTR. For example, Larish proposed a new predictive LTR that can provide a time-advanced measure of rollover propensity and, therefore, offers significant benefits for closed-loop rollover prevention [37]. And Li et al. introduced an improved predictive LTR (*IPLTR*) as the rollover index based on an 8-DOF nonlinear vehicle model [38].

3.2   ROLLOVER WARNING

The above-mentioned rollover indexes are effective to monitor the process of vehicle. However, it tends be too slow to actively prevent the rollover, especially for high CG vehicles with slow brake actuators.

3.2.1   TIME-TO-ROLLOVER

The Time to Rollover (TTR) is one of the most efficient indicators in order to anticipate the rollover detection. It is defined as the time remaining before wheel lift off will occur, which gives a clear indication of the beginning of rollover. Chen and Peng used TTR which is computed from yaw-roll model to prevent rollover for Sports Utility Vehicles [7]. Zhu et al. proposed a conventional time-to-rollover warning algorithm which was presented based on the 3-DOF vehicle model and used to investigate integrated chassis control to prevent vehicle rollover [39]. Dahmani et al. computed TTR by assuming that the LTR increases or decreases at its current rate in the near future, compute the time taken by the LTR to reach 1 or -1, and then compared the predictive effect of TTR and LTR. Considering that the vertical load of each wheel is difficult to be measured or estimated in real time, it can be transformed into other expression, as shown in Section 3.1.3. This is denoted by LTR_d. The computational formula of TTR is shown as follows [31, 40]:

$$TTR = \frac{1 - LTR_d}{R_{LTR}} \quad \text{if} \quad LTR > 0 \tag{3.10}$$

$$TTR = \frac{-1 - LTR_d}{R_{LTR}} \quad \text{if} \quad LTR < 0, \tag{3.11}$$

with R_{LTR} being the LTR rate, which is obtained from a filtered differential signal of the LTR.

In order to test the validity of TTR, two Fishhook tests are conducted with different steering wheel angles, as shown in Figure 3.13. The input steering angle used in test 2 is defined such that the wheel liftoff occurs at 2.8 s, whereas in test 1 no wheel liftoff occurs. In this simulation, the vehicle is driven at a constant speed of 110 km/h in a 6% banked road.

Figure 3.14 shows the comparison of TTR in the two tests. TTR shows good efficiency for the rollover detection, and the proposed rollover indicator, which is the TTR, bigger anticipation

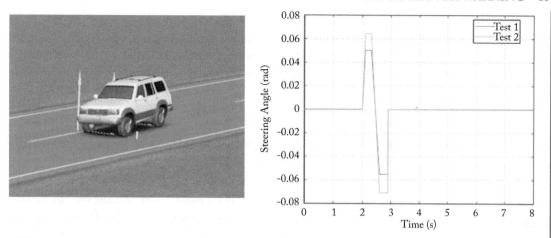

Figure 3.13: The used steering angles in test 1 and test 2.

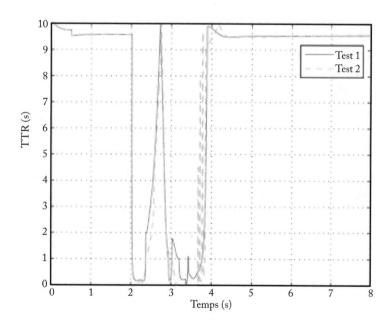

Figure 3.14: Comparison of TTR in test 1 and test 2.

in the rollover detection. This advantage is very interesting since the rollover must be avoided in a matter of seconds.

3.2.2   PREDICTION ROLLOVER WARNING

A new predictive LTR (PLTR) is developed by Chad [37] can provide a time-advanced measure of rollover propensity and, therefore, offers significant benefits for closed-loop rollover prevention. First, a common rollover index (LTR_e) based on the LTR is given as:

$$LTR_e = \frac{2h}{T_w g}\left[a_y + g \cdot \sin\phi\right], \tag{3.12}$$

where a_y is the measured lateral acceleration of the vehicle, h is the distance from sprung mass CG to roll center, T_w is the track width, and ϕ is the vehicle roll angle.

  With this index, the time between detection of potential rollover characteristics and the moment rollover occurs may sometimes be too small for a rollover prevention system to stop the vehicle from rolling over. Hence, a new predictive rollover index, i.e., PLTR, is developed. This predictive index indicates future vehicle rollover propensity for a wide range of vehicle maneuvers, based on data collected in the current time frame. The PLTR is defined as follows:

$$PLTR_{t_0}(\Delta t) = LTR(t_0) + L\dot{T}R(t_0) \cdot \Delta t, \tag{3.13}$$

where Δt is the preview time, and t_0 is the current time.

  Considering the LTR_e from Equation (3.12), we have

$$PLTR_{t_0}(\Delta t) = LTR(t_0) + \frac{2h}{T_w g}\left[a_y + g \cdot \sin\phi\right] \cdot \Delta t, \tag{3.14}$$

or

$$PLTR_{t_0}(\Delta t) = LTR(t_0) + \frac{2h}{T_w g}\left[a_y + g\phi\right] \cdot \Delta t. \tag{3.15}$$

  Equation (3.15) shows the calculation of the PLTR at time t_0 that is predicted for a future time horizon Δt. a_y is typically noisy, and it is difficult to obtain a smooth value of its derivative. A filtering technique is first used to address this problem, as shown in the following equation:

$$PLTR_{t_0}(\Delta t) = \frac{2h}{d}\left[\frac{a_y}{g} + \sin\phi\right] + \frac{2h}{T_w \cdot g}\left[\frac{s}{\tau s + 1}a_y + \frac{\tau s}{\tau s + 1}\dot{a}_y + g\phi\right] \cdot \Delta t, \tag{3.16}$$

where τ is the time constant.

  The lateral acceleration derivative in the second term can be further estimated from the lateral dynamics. By utilizing a linear approximation and the small angle assumption, the lateral dynamics equation can be written as:

$$ma_y = -C_0\beta - C_1\frac{r}{U} + 2C_f\delta, \tag{3.17}$$

where, $C_0 = 2C_f + 2C_r$ and $C_1 = 2aC_f - 2bC_r$. C_f and C_r are the cornering stiffness values for the front and rear tires, respectively. r is the yaw rate of the vehicle and U is the vehicle speed.

The derivative of Equation (3.17) can be written as

$$\dot{a}_y = \frac{-C_0\left(a_y - rU\right) - C_1\dot{r}}{mU} + \frac{2C_f}{m}\frac{1}{\tau_{sw}s + 1}\frac{1}{SR}\dot{\delta}_d,$$
(3.18)

where $(\delta/\delta_d) = (1/SR) \cdot (1/\tau_{sw}s + 1)$, δ_d is the driver's steering-wheel angle, τ_{sw} is the steering first-order time constant, and SR is the steering ratio.

By using this model-based filter, the noise from the differentiation of the steering-wheel angle can be filtered out using a low-pass filter. Moreover, the driver's steering input information plays an important role in predicting the rollover index due to the inherent time delay between the steering input and its influence on vehicle roll.

The new PLTR is displayed as follows:

$$PLTR_{t_0}(\Delta t) = \frac{2h}{T_w}\left[\frac{a_y(t_0)}{g} + \sin\phi\right] + \frac{2h}{T_w \cdot g}\left[\frac{s}{\tau s + 1}a_y(t_0)\right.$$
$$\left. + \frac{\tau s}{\tau s + 1}\left(\frac{-C_0\left(a_y - ru\right) - C_1\dot{r}}{mU} + \frac{2C_f}{m}\frac{1}{\tau_{sw}s + 1}\frac{1}{SR}\dot{\delta}_d\right) + g\dot{\phi}\right] \cdot \Delta t.$$
(3.19)

Filter $\tau s^2/((\tau s + 1)(\tau_{sw}s + 1))$ is used on the driver's steering angle. Prediction time Δt needs to be selected to be long enough to cover the rollover prevention system response time.

The term $\sin(\phi)$ is approximately proportional to lateral acceleration. Hence, $\sin(\phi)$ can be replaced by ka_y. The value of constant k depends on the CG height and suspension parameters and will have to be accordingly tuned for each vehicle. For small roll angles, the term can be entirely ignored. Finally, the final form of the new PLTR is given as follows:

$$PLTR_{t_0}(\Delta t) = \frac{2h}{T_w g}(1 + kg)a_y(t_0) + \frac{2h}{T_w \cdot g}\left[\frac{s}{\tau s + 1}a_y(t_0)\right.$$
$$\left. + \frac{\tau s}{\tau s + 1}\left(\frac{-C_0\left(a_y - ru\right) - C_1\dot{r}}{mU} + \frac{2C_f}{m}\frac{1}{\tau_{sw}s + 1}\frac{1}{SR}\dot{\delta}_d\right) + g\dot{\phi}\right] \cdot \Delta t.$$
(3.20)

The simulation was performed using a stock Humvee vehicle model in Carsim to illustrate the effectiveness of the PLTR with 0.3 s predictive time [37]. Figure 3.15 shows how the PLTR matches the actual LTR profile and the predictive quality of the PLTR. It is shown that the PLTR shows a time advance (of the order of 100 ms) compared with the LTR. Otherwise, the PLTR roughly matches the shape of the LTR trajectory.

Figure 3.16 shows the calculated LTR and PLTR from experimental vehicle test data for the "Sine with Dwell" maneuver (open loop). The plots show a good correlation between the simulation study and the actual implementation in the vehicle.

Figures 3.17 and 3.18 further present the calculation of the LTR and the PLTR from experimental vehicle testing data for the North Atlantic Treaty Organization double-lane change

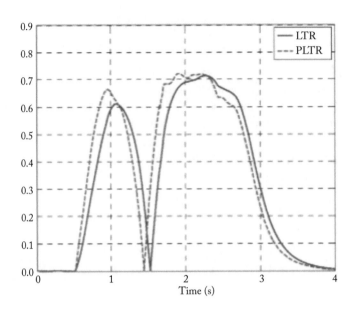

Figure 3.15: "Sine with Dwell" at a steering amplitude of 113.8° (simulation).

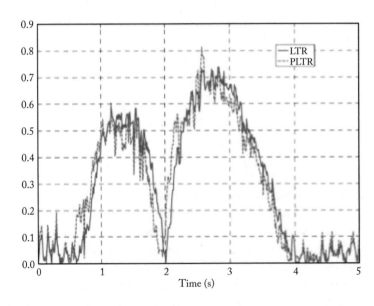

Figure 3.16: "Sine with Dwell" at a steering amplitude of 113.8° (experimental measurements).

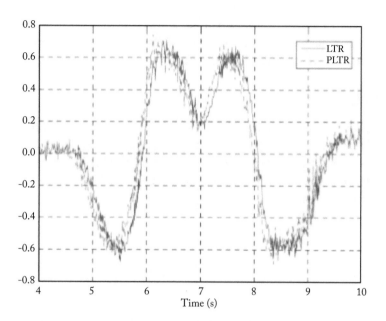

Figure 3.17: Double-lane change (experimental measurements).

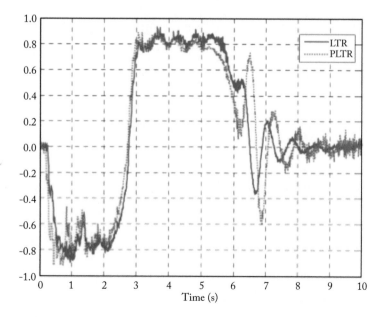

Figure 3.18: Fishhook maneuver (experimental measurements).

maneuver at 90 km/h and a Fishhook maneuver at 64 km/h, respectively. It is also obvious from these figures that the PLTR provides a small but important time advance compared with the LTR index. Further, in all these experimentally validated steering maneuvers, there is a good match in the shape of LTR and PLTR trajectories.

3.3   SUMMARY

As scholars develop a deeper understanding, the stability of untripped rollover is well documented. The rollover index of untripped rollover includes SSF, DSF, and LTR. The LTR is the most effective. In order to predict the vehicle rollover state, sufficient reaction time is left for the driver, scholars put forward the early warning method of rollover includes time-to-rollover and prediction rollover warning.

Stability of Tripped Vehicle Rollover

At present, most studies on the stability of rollover focus on untripped rollover. However, tripped rollover accounts for a large proportion of rollover accidents. It is necessary to analyze the tripped rollover deeply. Some scholars have put forward the roll index for tripped rollover on uneven roads, slopped roads and banked roads and so on. And the energy methods for the stability of tripped rollover is also analyzed.

4.1   ROLL INDEX OF TRIPPED VEHICLE ROLLOVER

4.1.1   ROLLOVER INDEX ON UNEVEN ROADS

According to the definition of the LTR and yaw-roll-vertical model mentioned in Section 2.4, an improved definition of the lateral load transfer ratio is proposed to describe the tripped vehicle rollover on uneven roads by Jin et al. [15]. From the dynamics of the wheel vertical motion, the total vertical load is consisted of the dynamic vertical force and the static vertical force of the tire. So, the total vertical loads of the left and right wheels can be described as follows:

$$F_{z1} = -k_{t1}\left(z_{u1} - z_{r1}\right) + mg/2 \tag{4.1}$$

$$F_{z2} = -k_{t2}\left(z_{u2} - z_{r2}\right) + mg/2. \tag{4.2}$$

Substituting Equations (4.1) and (4.2) in Equations (2.21) and (2.22), respectively, the total vertical loads will be rewritten as

$$F_{z1} = m_1\ddot{z}_{u1} + F_{s1} + mg/2 \tag{4.3}$$

$$F_{z2} = m_2\ddot{z}_{u2} + F_{s2} + mg/2. \tag{4.4}$$

From Equations (3.4), (4.3), and (4.4), a new index can be defined as follows to evaluate the danger of vehicle rollover for both tripped and untripped:

$$RI = \frac{m_1\ddot{z}_{u1} - m_2\ddot{z}_{u2} + F_{s1} - F_{s2}}{m_1\ddot{z}_{u1} + m_2\ddot{z}_{u2} + F_{s1} + F_{s2} + mg}. \tag{4.5}$$

As the suspension forces are difficult to obtain in real time, the rollover index in Equation (4.5) also cannot predict the risk of vehicle rollover directly. Fortunately, the difference between the left suspension force and the right suspension force can be obtained from Equation (2.19), and the sum of the left suspension force and the right suspension force can be obtained from Equation (2.20). That is,

$$F_{s2} - F_{s1} = -\frac{2}{T_w}\left[I_x\ddot{\phi} - m_s h a_y - m_s h g\phi\right] \qquad (4.6)$$

$$F_{s1} + F_{s2} = m_s\ddot{z}_c. \qquad (4.7)$$

Substituting Equations (4.6) and (4.7) in Equation (4.5), the rollover index for both untripped and special tripped situations can be rewritten as

$$RI = \frac{T_w\left(m_1\ddot{z}_{u1} - m_2\ddot{z}_{u2}\right) - 2\left(\left[I_x\ddot{\phi} - m_s h a_y - m_s h g\phi\right]\right)}{T_w\left(m_1\ddot{z}_{u1} + m_2\ddot{z}_{u2} + m_s\ddot{z}_c + mg\right)}. \qquad (4.8)$$

The new rollover index can be obtained in real time with some unknown parameters which can be measured and estimated, such as vertical accelerations of the sprung mass and unsprung mass, lateral acceleration and roll angle.

To verify the new rollover index, the dynamic performance of a vehicle rollover is simulated in CarSim, an industry-standard vehicle dynamic simulation software. The model of a large passenger vehicle is used, and the lateral load transfer ratio and new rollover index are obtained in three different cases below.

Case I

In an untripped rollover situation, the vehicle roll is induced by driver's maneuver, such as the step steering at 1.0 s. The final value of the steering angle of the front wheel is $\delta = 2°$, and the vehicle speed is 100 km/h. Then, the untripped rollover indices of the vehicle can be obtained.

As shown in Figure 4.1a, the new rollover index agrees with the fundamental definition of the LTR very well when the vehicle rolls with all wheels keeping on road due to step steering.

Case II

In a tripped rollover situation, the vehicle rollover happens due to external road input, such as an unpredictable road bump under the right wheel when vehicle moves on a straight lane. In this case, the maximum height of the road bump is 0.15 m, and the vehicle speed is 100 km/h. The tripped rollover indices of the vehicle are plotted in Figure 4.1b.

The improved rollover index also fits the fundamental definition of the lateral load transfer ratio very well in a tripped rollover situation, except that the right wheels lift off a few seconds. So, the new rollover index can be used to detect the danger of tripped rollover if its value is outside the range from -1 to 1.

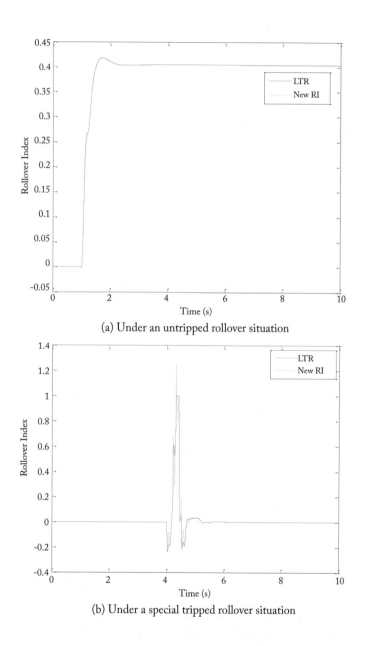

(a) Under an untripped rollover situation

(b) Under a special tripped rollover situation

Figure 4.1: Rollover indices under different situations. (*Continues.*)

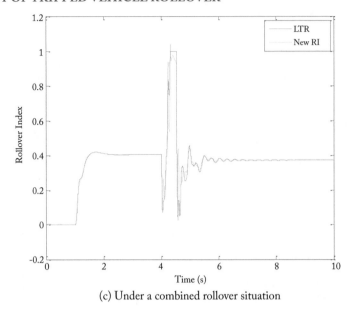

(c) Under a combined rollover situation

Figure 4.1: (*Continued.*) Rollover indices under different situations.

Case III

In this case a combined untripped and tripped rollover due to an unpredictable road bump under the right wheel while driving in a step steering is studied. The final value of the steering angle of the front wheel is $\delta = 2°$, the maximum height of the road bump is 0.15 m, and the vehicle speed is 100 km/h. The rollover indices of the vehicle in this case is shown in Figure 4.1c.

The same conclusions can be drawn as the above two cases. So, the improved rollover index can predict the risk for both tripped and untripped rollover of a vehicle by measuring the accelerations and roll angle in real time.

4.1.2   ROLLOVER INDEX ON BANKED ROADS

Much progress has been made for the bus rollover warning in the past decades. But much of this research has not taken account for dynamic road bank. To fill in the gap, some researchers present real-time rollover trend prediction to indicate bus rollover risk with road bank estimation [4, 5, 36, 40, 41]. Dahmani et al. used the estimated roll angle and roll rate to compute the rollover index which is based on the prediction of the lateral load transfer [31]. Lateral load transfer is the change in the normal force acting on the tyres due to both the acceleration of the CG, and the shifting of the position of the CG in the y direction due to the movement of the suspension. Figure 4.2 illustrates lateral load transfer in the vertical plane.

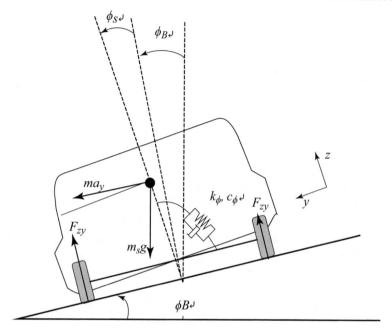

Figure 4.2: Vehicle roll model.

The estimation of the LTR is very difficult since normal force sensors are expensive. An expression for LTR which depends on the roll states and vehicle parameters can be obtained. This is denoted by LTR$_d$. In order to derive LTR$_d$, we resolve weight ($m_s g$) and pseudo-force ($m_s a_y$) into components in the vehicle-fixed y and z directions. The following dynamics are obtained:

$$m_s a_y h + m_s g h (\phi_s + \phi_B) - c_\phi \dot{\phi}_s + k_\phi \phi_s = 0, \tag{4.9}$$

where m_s is the sprung mass of vehicle and a_y is the lateral acceleration. k_ϕ and c_ϕ are the combined roll stiffness coefficient and combined roll damping coefficient, respectively. h is the CG height from roll axis.

The torque balance for the sprung and unsprung masses about the left tyre roll axis is written as follows:

$$F_{z2} T_w + m_s a_y h + m_s g h \phi_B - m_s g \left(\frac{T_w}{2} - h\phi_s \right) - m_u g \frac{T_w}{2} = 0, \tag{4.10}$$

where F_{z2} are the vertical right tyre forces, m_u is the unsprung mass of vehicle.

By substituting Equations (4.9) and (4.10) into Equation (3.4), the following expression for LTR$_d$ can be obtained:

$$LTR_d = \frac{2}{mgT_w}\left(c_\phi \dot{\phi}_s + k_\phi \phi_s\right).$$

(4.11)

4.2   ENERGY METHODS

Besides the methods mention above, there have been some efforts to predict vehicle rollover using the vehicle roll energy and the rollover potential energy. Choi proposed a new rollover index using energy method [42]. Figure 4.3 shows the diagram of the front view of a vehicle sprung mass where y and z axes are fixed to the CG of the sprung mass and rotate with the mass. a_y is the lateral acceleration measured by the accelerometer attached to the vehicle sprung mass. The measured acceleration is partly from the vehicle acceleration and partly from gravity. ϕ is the absolute roll angle of the vehicle sprung mass with respect to the earth coordinates due to the lateral acceleration and/or the super elevation angle of the road surface.

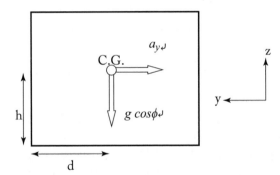

Figure 4.3: Diagram of vehicle coordinates.

If the artificial angle σ is defined by the tangent ratio of a_y and $g\cos(\phi)$ as follows:

$$\tan\sigma \equiv \frac{a_y}{g\cos\phi}.$$

(4.12)

Then, Figure 4.3 can be reconfigured as Figure 4.4.

In the reconfigured coordinates, z axis is defined as parallel to the direction of the net force on the vehicle sprung mass. Defining the net acceleration $g\cos(\phi)/\cos(\sigma)$ on the vehicle mass as virtual gravity, the problem can be reduced to a mass on a σ degree hill with a gravity constant of $g\cos(\phi)/\cos(\sigma)$.

In Figure 4.4, the current height of CG is $d\sin(\sigma) + h\cos(\sigma)$, and the critical height of CG, where the vehicle is at the verge of rollover, is $\sqrt{d^2 + h^2}$. Therefore, defining the height

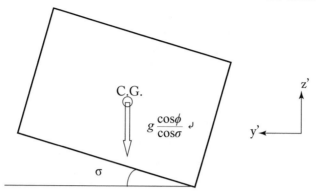

Figure 4.4: Diagram of a vehicle in virtual gravity coordinates.

change of CG required for rollover as Δh:

$$
\begin{aligned}
\Delta h &= \sqrt{d^2 + h^2} - (d \sin \sigma + h \cos \sigma) \\
&= \sqrt{d^2 + h^2} - \frac{d a_y + h g \cos \phi}{\sqrt{g^2 \cos^2 \phi + a_y^2}}.
\end{aligned}
\tag{4.13}
$$

The minimum amount of potential energy—normalized with the vehicle mass—required for the rollover is defined as $(g \cos \phi / \cos \sigma) * \Delta h$ using the concept of virtual gravity constant. Since the lateral kinetic energy of a vehicle can be converted to potential energy very quickly through the roll motion, a vehicle has the potential to rollover as long as the lateral energy is larger or equal to the minimum required potential energy, i.e.,

$$
\frac{1}{2} v^2 > \frac{g \cos \phi}{\cos \sigma} \Delta h = \sqrt{g^2 \cos^2 \phi + a_y^2} \sqrt{d^2 + h^2} - (d a_y + h g \cos \phi).
\tag{4.14}
$$

The lateral velocity v can be calculated from longitudinal velocity u and vehicle side slip angle β as

$$
v = U\beta.
\tag{4.15}
$$

Motivated by the above inequality condition, a rollover potentiality index Φ_0 is defined as follows:

$$
\Phi_0 = \frac{1}{2} |U\beta|^2 - \sqrt{g^2 + a_y^2} \sqrt{d^2 + h^2} + d a_y + h g.
\tag{4.16}
$$

Positive Φ_0 means that the vehicle has the potential to rollover, and the possibility of rollover increases with Φ_0. However, large Φ_0 alone does not mean that the vehicle will rollover. The large kinetic energy needs to be converted to roll dynamic energy. It usually happens when a vehicle hits a high μ surface or a bump after a large side slip typically on a low μ surface. If the vehicle hits a high μ surface, the lateral acceleration of the vehicle increases very quickly. Simulation results show that the measured lateral acceleration needs to be more than 80% of statically critical lateral acceleration for rollover to happen. However, this 80% acceleration threshold (A_{Th}) needs to be tuned for different vehicles with different vehicle parameters and suspension characteristics. Statically, critical lateral acceleration is defined as the acceleration to make a vehicle to rollover on a flat surface and described as $(d/h)g$.

Φ_0 and measured lateral acceleration, the rollover index (RI) is defined as follows:

$$\Phi = \Phi_0 \times \left(|a_y| - \frac{d}{h}g \times A_{Th} > 0 \right). \tag{4.17}$$

It can be seen that the RI can also describe the stability of tripped rollover from its derivation process.

4.3   SUMMARY

In conclusion, some efforts has been made to analyze actively control tripped rollover in some special cases such as uneven roads and banked roads. However, there are many reasons for tripped rollover, and the corresponding stress state of the vehicle is also very complex, it is very difficult to give early warning and control of tripped rollover. In the future, we need to focus more research on tripped rollover.

CHAPTER 5

Active Control for Vehicle Rollover Avoidance

Rollover avoidance control is another important subject for the study of rollover stability. With the development of advanced control technology and cost reduction of electronic and control equipment, various active control systems have been widely used in the automotive industry in the design of anti-rollover control systems. At present, five main actuators have been proposed and widely studied to prevent vehicle rollover: active anti-roll bar, active suspension, differential braking, active steering, and combinations of these different techniques. Based on these actuators, many researchers proposed advanced control algorithm to realize the vehicle's anti-rollover.

5.1 ANTI-ROLL BAR SYSTEM

The anti-roll bar can adjust its deformation according to the vehicle motion by adding hydraulic or electric actuator, and thus to control the posture of vehicle body. As shown in Figure 5.1, a stiff U-shaped anti-roll bar is connected to the trailing arms directly and to the vehicle frame by a pair of double-acting hydraulic actuators. The position of the anti-roll bar is therefore determined by both the wheel positions and the actuator positions. By extending one actuator and retracting the other, it is possible to apply a roll moment to the sprung mass and tilt the vehicle body. First, Electronic Control Unit (ECU) collects relevant sensor signals including steering wheel angle sensor, vehicle speed sensor, vehicle roll angle sensor, etc. According to control algorithm, ECU calculates required roll moment of anti-roll bar, then the displacement of actuators can also be determined. ECU controls the elongation or contraction of the actuators on both sides through the drive circuit, the corresponding anti-roll moment is generated to prevent vehicle rollover [43].

Many control algorithms have been proposed to determine roll moment of active anti-roll bar. Huang et al. [2] and Vu et al. [44] present the application of Linear Quadratic Regulator (LQR) algorithm for rollover prevention of heavy articulated vehicles with active anti-roll bar control. Muniandy investigated a self-tuning fuzzy proportional-integral -derivative controller to for active anti-roll bar [45]. H∞ approach [46] and sliding-mode control method [28] are also used to design active anti-roll bars.

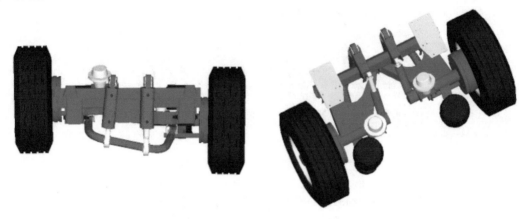

Figure 5.1: Active anti-roll bar general arrangement.

5.2   ACTIVE SUSPENSION SYSTEM

The basic idea is that the equivalent roll damping and equivalent roll stiffness of active suspension can be adjusted. Therefore, the handing stability and ride performance are improved. The schematic diagram of active suspension is shown in Figure 5.2. Active suspension should have three conditions: power producer which can generate acting force; actuating element which can transfer the acting force; and sensors for collecting data and ECU for operation. When the vehicle loads, speed, road condition, and so on, changes the sensors collect the status signal of vehicle and feed back to ECU. ECU controls the power producer to generate the corresponding acting force, the damping and stiffness of suspension can be adjusted automatically. The equation of roll motion of a vehicle with an active suspension is shown as follows:

$$\ddot{\phi} = \frac{1}{I_x} \left[-c_\phi \dot{\phi} - \left(k_\phi - mgh \right) \phi + ma_y h + M_{RC} \right] + \Delta. \tag{5.1}$$

Among the intelligent safety technologies for road vehicles, active suspensions controlled by embedded computing elements for preventing rollover have received a lot of attention. Sarel et al. presented the possibility of using slow active suspension control to reduce the body roll and thus reduce the rollover propensity [47]. Zhu and Ayalew focuses on the application of active suspensions to vehicles with solid-axles for medium and light duty trucks [48]. The active suspension is also used in heavy-duty vehicles [33]. Active suspension can effectively resolve the contradictions between vehicle ride comfort and stability. However, a new contradiction between the active suspension performance and efficiency is aroused. Active suspension with excellent performance requires high actuation power and force in an aggressive condition, which is usually an excessive capacity for normal conditions. Sun et al. investigated on the efficiency and utilization rate of vehicle active suspension based on a 7-DOF full vehicle mode with a

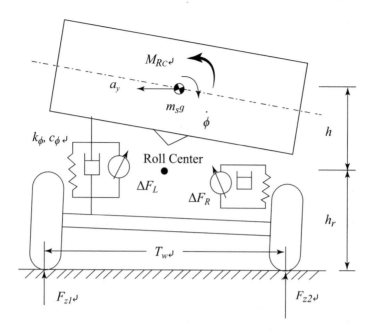

Figure 5.2: Schematic diagram of active suspension system.

linear quadratic Gaussian active suspension controller to improve the efficiency and capacity utilization rate [49].

5.3 ACTIVE STEERING SYSTEM

Active steering systems could reduce or reverse the unstable roll maneuver by controlling the steering angle. Active steering control is integrated with the steering system with a set of gearboxes that adjust steering drive according to speed. The system includes a fist-sized planetary gear and two input shafts. One input shaft connected to the steering wheel, the other controlled by an electric motor. As shown in Figure 5.3, the active steering can give a small superimposed steering angle in addition to the driver's given steering angle. When the vehicle is in danger of rollover because of a large steering angle, the active steering system automatically adjusts the steering angle to prevent vehicle rollover.

Active steering control can increase the turning radius, reduce the yaw motion, and effectively improve the stability of anti-rollover. So, many researchers have used active steering system as the actuator including active front steering control [32, 50–52], active rear pulse steering control, and four-wheel steering.

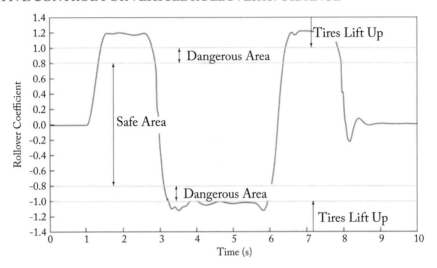

Figure 5.3: Rollover coefficient estimator for a double-lane change maneuver.

5.3.1   ACTIVE FRONT STEERING CONTROL

The active front steering control is to increase or decrease the steering angle of the front wheel under the premise of the steering angle controlled by the driver according to the driving state of the vehicle, thus improving the maneuverability and stability of the vehicle.

5.3.2   PULSE ACTIVE REAR STEERING

The pulse active rear steering (PARS) system focuses on reducing the likelihood of vehicle rollover by sending steering pulses to a single rear wheel whenever a rollover coefficient is outside a predetermined range. The system provides extra steering with respect to a current steering input from the driver until the rollover coefficient ultimately drops into safe area, as shown in Figure 5.3.

Table 5.1 shows the key differences between the conventional active rear steering and PARS. As can be seen, the concept of applying a series of pulses is similar to the standard commercial anti-lock braking system (ABS). The PARS system is a simple, inexpensive, and lightweight component that can be added to the vehicle when the vehicle already has ABS. Sensors for ABS can be simultaneously used to obtain the various measurements for this system. In addition, the system can be used in conjunction with other stability or yaw control or rollover control systems such as differential braking systems. The PARS system consists of two main parts: a pulse-generating device and a switch controller. The generated pulses provide the controller's steering input which, together with the driver's steering input, helps to control the vehicle's rollover. The switch controller includes a rollover estimator that calculates the rollover coefficient of the vehicle; it is only activated when the potential for rollover is detected. Zhang

Table 5.1: Comparison between conventional active rear steering and pulse active rear steering

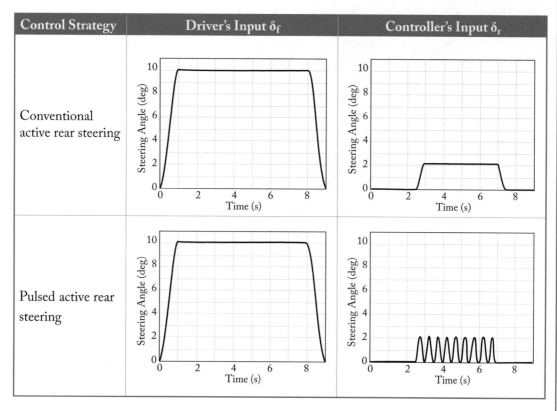

Control Strategy	Driver's Input δ_f	Controller's Input δ_r
Conventional active rear steering		
Pulsed active rear steering		

et al. proposed a novel pulse active rear steering system (PARS) for improving vehicle yaw stability [53], as shown in Figure 5.4. The hydraulic-mechanical pulsed steering system is designed and a model that represents the pulsed steering system characteristics is introduced. The oil volume sent to the hydraulic cylinders is proportional to the angle of the rotary steering valve, which is driven by a motor, shown schematically in Figure 5.4b. The steering system can be represented by a combination of a steering sliding valve and a hydraulic cylinder ram, illustrated in Figure 5.4b.

5.3.3   FOUR-WHEEL STEERING

Four-wheel steering refers to that four wheels can simultaneously deflect from the vehicle body according to the signal of the front wheel or the vehicle speed during vehicle steering. When the steering wheel angle is small, the rear wheel and the front wheel turn in the same direction, this reduces the turning radius of the vehicle. When the steering wheel angle is large, the rear wheel and the front wheel turn in the reverse direction, thus the handing stability is improved. Four-

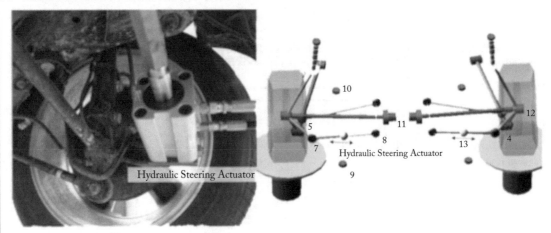

(a) The implementation of PARS

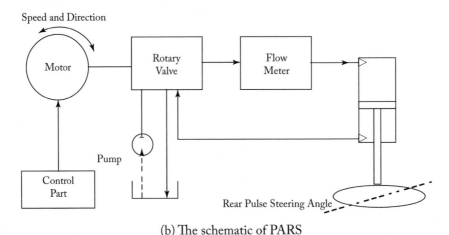

(b) The schematic of PARS

Figure 5.4: The implementation on a multi-link suspension and the schematic of PARS.

wheel steering can basically maintain the vehicle mass center side-slip angle is zero in the process of steering and can improve the dynamic response characteristic of the vehicle to steering wheel input. To a certain extent, the transient response performance indexes of the yaw angular velocity and lateral acceleration are improved, and the stability of the vehicle at high speed is obviously improved. At high speed, four-wheel steering system through the same direction of the rear and front-wheel steering, can effectively reduce/eliminate the vehicle sideslip, and improve vehicle stability and safety. Ono et al. used 4-wheel-distributed steering and 4-wheel-distributed traction/braking systems to improve the handing stability of vehicle [54]. The hierarchical control structure shown in Figure 5.5 is adopted for vehicle dynamics control.

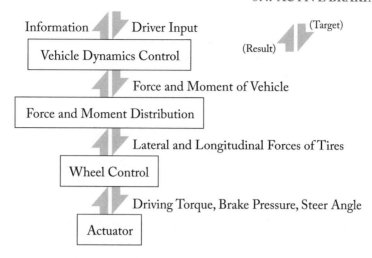

Figure 5.5: Hierarchical vehicle dynamics management algorithm.

Vehicle Dynamics Control: this layer calculates the target force and moment of the vehicle to achieve a desirable vehicle motion corresponding to the driver's pedal input and steering wheel angle. The determined target resultant force and moment also satisfy the robust stability condition.

Force and Moment Distribution: the target resultant force and moment of the vehicle motion are distributed to target tire forces of each wheel based on the friction circle of each wheel in this layer.

Wheel Control: this layer controls each wheel motion to achieve the target tire force.

5.4   ACTIVE BRAKING SYSTEM

Differential braking system applies unequal braking force to the different wheels, which generates a corresponding additional yawing moment to change the yaw rate of the vehicle. On the other hand, it can also decrease the vehicle speed, thus the anti-rollover property of vehicle is improved. Some researchers used differential braking to keep the value of *LTR* below a certain level and yield robustness to variations in vehicle speed [29, 55]. More control effort would be exerted for drivers with poor driving skills, and vice versa. Zhu proposed a sliding mode control (SMC)-based differential braking controller which based on a novel driver-adaptive Vehicle Stability Control (DAVSC) strategy [20].

Figure 5.6 shows the yaw moment change when the braking force is applied at front-inner wheel, rear-inner wheel, front-outer wheel, and rear-outer wheel, respectively. It is observed that applying the braking force to the front-outer wheel or rear-inner wheel can better prevent the vehicle from yaw motion, thus improving the roll stability of the vehicle.

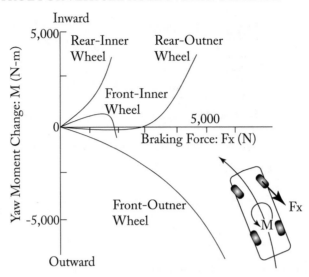

Figure 5.6: Yaw moment change.

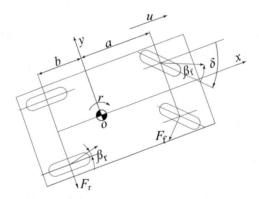

Figure 5.7: Bicycle vehicle model.

Although the actuator can provide enough brake torque, its magnitude at each wheel is still limited by some other factors, such as the coefficient of road adhesion and tires vertical load. From the bicycle vehicle model shown in Figure 5.7, the brake force at each wheel is constrained by:

$$F_{b2} \leq \frac{\mu mg(1 + LTR)}{2(a + b)},$$

(5.2)

where μ is the coefficient of road adhesion, and F_b is the brake force of front-outer wheel.

The yaw torque can be then expressed as:

$$M_B = \frac{T_w}{2}(F_{b1} - F_{b2}).$$ (5.3)

As known in the above section, differential braking is one of the main techniques for vehicle rollover avoidance. Also, Electro Hydraulic Brake (EHB) systems can produce enough brake torques fast enough for rollover prevention. The model and dynamic characteristics of an EHB was studied theoretically and mathematically in Yong et al.'s previous research [56].

Figure 5.8 shows a hydraulic scheme of electro hydraulic brake system including two different electro hydraulic valves. The first are controlled by Pulse Width Modulation (PWM) from the Electronic Control Unit (ECU), including inlet valves and outlet valves. The second are on/off valves, such as balance valves and cut valves. All the valves are 2/2 solenoid valve.

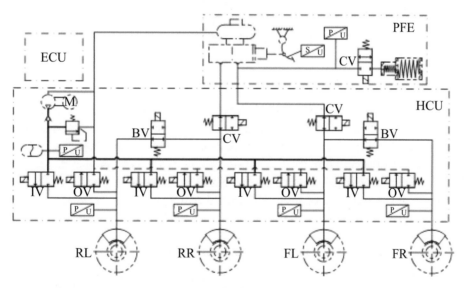

Figure 5.8: Hydraulic scheme of Electro Hydraulic Brake system.

When the driver brakes, in normal operation, the stroke sensor unit receives the signal of driver's brake intention and sends it to ECU. For the ECU switching the solenoid valves into different state. There are three operation modes of EHB system.

(1) Pressure build-up. The inlet valve and cut valve act in opposition state to each other, the inlet valve is open and the cut valve closed. Since the outlet valve holds initial state closed, the wheel brake cylinder is connected to a high-pressure accumulator fed by a pump. The brake fluid flows out of the high-pressure accumulator into the wheel brake cylinder and the brake pressure increases. (2) Pressure holding. Depending on requirements, as soon as enough pressure has been obtained in the wheel brake cylinder, the solenoid valves are switched to pressure holding state. In this state, all three valves are closed with respect to each other. The connections from

the wheel brake cylinder to the high-pressure accumulator, to the master cylinder, and to the reservoir are interrupted and the pressure of wheel brake cylinder is kept constant. (3) Pressure reduction. The outlet valve and cut valve act in opposition state to each other. The cut valve and the inlet valve closed, and the outlet valve is open. So, the wheel brake cylinder is connected to the reservoir. The brake fluid flows out of the wheel brake cylinder into the reservoir via the return line and the brake pressure drops.

It is necessary to include a failsafe operation for EHB system which contained many electronic elements, like sensors and solenoid valves. In this operation, all the valves hold initial state. The inlet valve and the outlet valve are closed and the cut valve holds open. So, the wheel brake cylinder is connected to the master cylinder, like the conventional hydraulic brake system.

The dynamic equations of the mode of pressure build-up and pressure reduction can be expressed as

$$\frac{dp_w}{dt} = \frac{C_b A_b}{K_b V_b} \left[p_0 \left(\frac{V_0}{V_0 + qt} \right)^\gamma - p_w \right]^n \tag{5.4}$$

$$\frac{dp_w}{dt} = -\frac{C_b A_b}{K_b V_b} p_w^n, \tag{5.5}$$

where C_b is the flow coefficient of solenoid valve; A_b is the area of orifice throttle; K_b is the bulk modulus of the brake oil; V_b denotes volume of the wheel brake cylinder; n is the index of solenoid valve; p_w is the pressure of wheel brake cylinder; q is average flow rate; γ is adiabatic index; and p_0 and V_0 denotes initial pressure and volume of the gas chamber of accumulator.

To valid the model of EHB system, a test platform is developed, as shown in Figure 5.9. The parameters of the model have been determined by means of the regressive analytics. The experimental results of the model parameters for each wheel are shown in Table 5.2.

Table 5.2: The experimental results of EHB model parameters

Parameter	Wheel 1	Wheel 2	Wheel 3	Wheel 4
n	0.83	0.83	0.85	0.85
$C_b A_b/(K_b V_b)$	28.2	28.2	58.5	58.5
q/ml/s	4.68	4.68	4.55	4.55

5.5   INTEGRATED CHASSIS SYSTEM

The four rollover avoidance control mechanisms mentioned above can effectively improve the stability of vehicle rollover. However, any single use of these mechanisms will have a certain adverse impact on the normal operation of the vehicle. For example, active steering systems could reduce or reverse the unstable roll maneuver by controlling the steering angle but change

Figure 5.9: The test platform of Electro Hydraulic Brake system.

the intention of driver operation; and different braking could reduce vehicle speed and use more energy. Therefore, to optimize the performance of rollover prevention control, combinations of these different techniques will be a better option. Yim et al. designed a controller that used an active anti-roll bar and an electronic stability program (ESP) for rollover prevention [57], and to enhance robustness of the controller, differential braking and an active suspension system were adopted as actuators that generated yaw and roll moments, respectively [8]. Doumiati et al. aimed at stabilizing the vehicle while achieving a desired yaw rate and proposed a suitable gain scheduled Linear Parameter Varying control strategy via coordination of active front steering and rear braking [58]. Moreover, Yoon et al. described a unified chassis control scheme which aimed to prevent vehicle rollover, and to improve vehicle maneuverability and its lateral stability by integrating electronic stability control (ESC) and active front steering (AFS). And the control scheme was evaluated on a virtual test track [59].

 Liu describes an integrated chassis control framework for a novel three-axle electric bus with active rear steering (ARS) axle and four motors at the middle and rear wheels. The integrated chassis control framework for the targeted electric bus is shown in Figure 5.10. The integrated framework consists of four parts: (1) an active speed limiting controller is designed for anti-body slip control and rollover prevention; (2) an ARS controller is designed for coordinating the tyre wear between the driving wheels; (3) an inter-axle torque distribution controller is designed for optimal torque distribution between the axles, considering anti-wheel slip and battery power limitations; and (4) a data acquisition and estimation module for collecting the measured and estimated vehicle states [60].

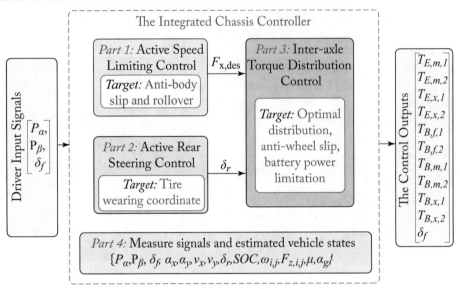

Figure 5.10: The framework of integrated chassis controller.

5.6   SUMMARY

In this chapter, the widely used rollover avoidance control methods are summarized and analyzed. The anti-roll bar, active steering system, differential braking, and active suspension all can work to prevent the vehicle rollover to some extent although each of them has its limitation. Therefore, the integrated chassis control will be the promising research direction for rollover avoidance in the future.

CHAPTER 6

Rollover Control Strategies and Algorithms

Effective control strategy and algorithm is the core of preventing vehicle rollover accidents. During the development of vehicle active rollover avoidance, many rollover control strategies and algorithms have been proposed by researchers. In the following section, some common control methods used in the literature are introduced and analyzed, such as Proportional-integral-derivative (PID) control, linear quadratic regulator (LQR) control, H-infinity control, and Model predictive control (MPC).

6.1 PROPORTIONAL-INTEGRAL-DERIVATIVE CONTROL METHOD

PID control is one of the most commonly used controllers, which has a wide range of applications. Regardless of whether the structure or parameters of the controlled plant are determined and accurate, PID control can play its own control effect by adjusting the parameters. The basic idea of PID control is that, based on the system variable error $e(t)$ (being the error between the expected value and actual value of feedback variables), the control variable is calculated through selecting the proper control gains and applied to controlled plant to reduce the error. The control law is shown in Equation (6.1):

$$u(t) = K_p \left[e(t) + \frac{1}{T_i} \int e(t)dt + T_d \frac{de(t)}{dt} \right]. \tag{6.1}$$

Figure 6.1 shows the control effect of PID controller. It is observed that PID control can prevent the occurrence of the vehicle rollover. However, the parameter of PID controller is difficult to be tuned in real time. So researchers usually use PID in combination with other control algorithms.

Muniandy proposed a PI–PD-type (proportional-integral–proportional-derivative) fuzzy controller to prevent vehicle rollover [45]. This controller added the fuzzy control into PI-PD-type controller. Figure 6.2 shows the basic layout of PI–PD controller.

The controller is designed to preserve the linear structure of conventional PI–PD controller and substitute the coefficient gains with nonlinear fuzzy functions. The output of PI–

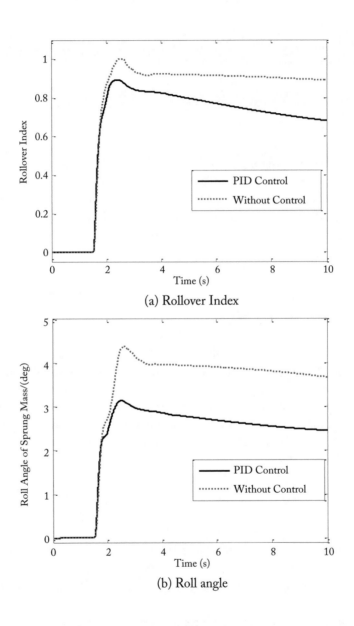

(a) Rollover Index

(b) Roll angle

Figure 6.1: Control effect of PID control. (*Continues.*)

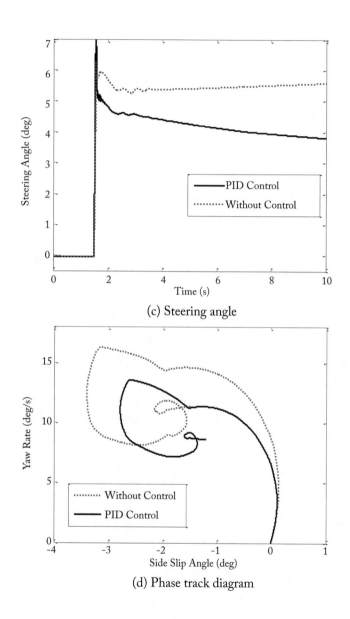

(c) Steering angle

(d) Phase track diagram

Figure 6.1: (*Continued.*) Control effect of PID control.

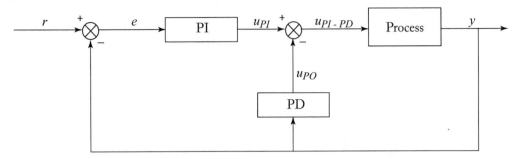

Figure 6.2: Basic layout of PI–PD controller.

PD-type fuzzy controller, $u_{PI-PD}(nT)$, is represented by

$$u_{PI-PD}(nT) = u_{PI}(nT) - u_{PD}(nT),\tag{6.2}$$

where $u_{PI}(nT)$ and $u_{PD}(nT)$ are the equivalent outputs from fuzzy PI and fuzzy PD controllers, respectively. Prior to that, in Laplace domain, both conventional analogue PI and PD controllers can be represented by

$$\begin{cases} u_{PI}(s) = \left(K_p^c + \dfrac{K_i^c}{s}\right) E(s) \\ u_{PD}(s) = K_p^{c'} + K_d^c Y(s), \end{cases}\tag{6.3}$$

where $u_{PI}(s)$ and $u_{PD}(s)$ are outputs of analogue PI and PD controllers, respectively; K_p^c, K_i^c, and K_d^c are proportional, integral, and derivative gain, respectively. It can be seen that PI controller is influenced by error signal $E(s)$ and PD controller is influenced by process output $Y(s)$. By applying bilinear transformation, Equations (6.3) is transformed into a discrete version. Hence, fuzzy PI controller output is written as

$$u_{PI}(nT) = u_{PI}(nT - T) + K_{uPI}\Delta u_{PI}(nT)\tag{6.4}$$

$$u_{PD}(nT) = -u_{PD}(nT - T) + K_{uPD}\Delta u_{PD}(nT).\tag{6.5}$$

By inserting Equations (6.4) and (6.5) into Equation (6.2), the output of PI–PD-type FLC controller will be

$$u_{PI-PD}(nT) = u_{PI}(nT - T) + K_{uPI}\Delta u_{PI}(nT) + u_{PD}(nT - T) + K_{uPD}\Delta u_{PD}(nT).\tag{6.6}$$

Both K_{uPI} and K_{uPD} will be determined by fuzzy rules. This controller's layout applied in active anti-roll-bar (ARB) is presented in Figure 6.3.

Similar to a standard fuzzy controller, membership functions and rules will be applied to the fuzzy PI and fuzzy PD controllers. The inputs for both controllers will be roll angle error signal. As the derivative controller receives roll angle signal directly from the system feedback itself, it is expected that the derivative kick phenomenon can be avoided. Earlier, it has been

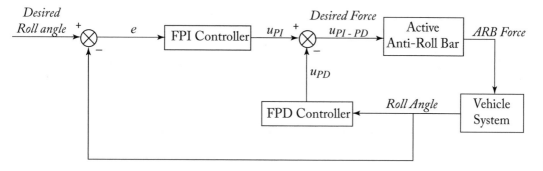

Figure 6.3: PI–PD-type fuzzy logic controller layout.

stated that fuzzy PI controller has two inputs, which are roll angle error signal, $e_p(nT)$ and rate of change of roll angle error signal, $e_v(nT)$. The fuzzy PI controller output (also called as incremental control output) is denoted as $\Delta u_{PI}(nT)$. The inputs for fuzzy PD controllers are the roll angle, $d(nT)$ (desired roll angle) and rate of change of roll angle, $\Delta y(nT)$. Figure 6.4a,b show the input membership functions for both PI and PD controllers. The input unit in Figure 6.4a is in degree, while the input unit in Figure 6.4b is °/s.

Figure 6.5 shows the output membership functions for both PI and PD controllers since both outputs are represented by the same membership functions. The output unit would be Newtons as it represents the required force by the actuator. The range of each membership functions is determined by typical operating range of a passenger car in a real application. The maximum value for required force is bound by the hardware capability.

A set of control rules base is created for fuzzy PI control as follows.

- RULE 1: IF e_p negative AND e_v negative, THEN PI-output = output negative.

- RULE 2: IF e_p negative AND e_v positive, THEN PI-output = output zero.

- RULE 3: IF e_p positive AND e_v negative, THEN PI-output = output zero.

- RULE 4: IF e_p positive AND e_v positive, THEN PI-output = output positive.

The membership functions have been kept simple in triangular form to reduce computing memory usage. The structure of the membership functions for both input signals are the same to avoid further memory allocations for the controller. The output signal for fuzzy PD controller is denoted as $\Delta u_{PD}(nT)$.

Fuzzy PD controller's rules set is as follows.

- RULE 5: IF d positive AND Δy positive, THEN PD-output = output zero.

- RULE 6: IF d positive AND Δy negative, THEN PD-output = output positive.

- RULE 7: IF d negative AND Δy positive, THEN PD-output = output negative.

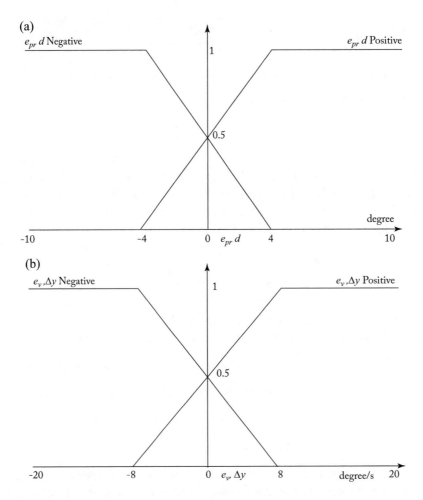

Figure 6.4: Membership functions: (a) for roll angle error (e_p) and average change of roll angle (d) and (b) for rate of change of roll angle error (e_v) and roll rate (Δy).

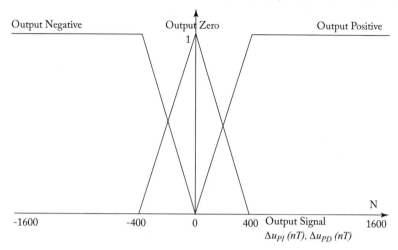

Figure 6.5: Membership functions for fuzzy PD and fuzzy PI output signals.

- RULE 8: IF d negative AND Δy negative, THEN PD-output = output zero.

Defuzzification process is done based on "center of mass" approach. To optimize the controller's performance, a set of parameters for the controller was obtained using trial and error method. It is an initial step to determine the test parameters where the parameters were chosen one by one while the changes occurring when the output were tracked. Individual values often increase gradually until the desired plant output response is achieved. Excessive amounts of actuator force and speed demands will also become the limiting factor for this method. After extensive simulation tests, parameter values satisfying all simulation tests are as follows: $Kp = 0.3140$, $Ki = 0.0971$, $K'p = 0.0576$, $K_d = 0.001$, $K_{uPD} = 0.4968$, and $K_{uPI} = 0.4496$. The sampling period is set to $T = 0.01$ second to cope with the vehicle suspension system response. Integral windup problem is not expected from this simulation because the simulation was executed in an ideal system assuming there is no saturation or physical limitation on the actuators and other related hardware. In future, when actual hardware implemented in this system, limiting the controller output according to the physical limit of the actual actuator is strongly recommended.

6.2   H-INFINITY CONTROL METHOD

To solve the rollover problem due to complex road conditions, the variation of the number of passengers, and other external interference, H-infinity control performance is used to provide robustness to model uncertainty and external disturbances [28, 31, 41, 46, 61–63]. Jin [15] designed a H∞ controller with differential braking as the actuator to prevent vehicle rollover and it is optimized by genetic algorithm.

As shown in Figure 6.6, e is the rollover index error, and r_{in} is the reference input of the rollover index, z_1, z_2, and z_3 are the evaluation outputs which dependent on exogenous input, G_s is the transfer functions which can be obtained using the state space equation of vehicle model mentioned in Section 6.2, K_c is the transfer function of H-infinity controller, w_d is the system input, RI is the rollover index, and M_B is the corrected yaw torque.

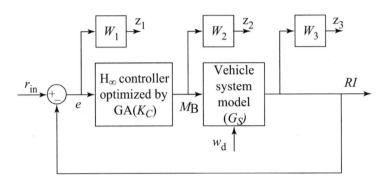

Figure 6.6: Block diagram of H-infinity controller for vehicle rollover prevention.

In order to formulate the standard structure of H-infinity controller, the weight functions W_1, W_2, and W_3 are defined to characterize the performance objectives and the actuator limitations.

W_1 weights the rollover index error signal. It is a constraint of the robustness of the rollover prevention control system which can adjust the influence of external interference.

W_2 weights the yaw moment signal. It is a constraint of the amplitude of anti-yaw torque due to differential braking.

W_3 weights the rollover index signal. It is a constraint of the stability of the rollover prevention control system. It also restricts the yaw rate and the vehicle lateral velocity evolution.

From the robust H-infinity theory, if $w_d = 0$, some transfer functions from input to outputs could be defined as:

$$G_1 \stackrel{def}{=} \frac{e(s)}{r_{in}(s)} = (I + G_s K_c)^{-1} \tag{6.7}$$

$$G_2 \stackrel{def}{=} \frac{M_B(s)}{r_{in}(s)} = K_c (I + G_s K_c)^{-1} \tag{6.8}$$

$$G_3 \stackrel{def}{=} \frac{RI(s)}{r_{in}(s)} = G_s K_c (I + G_s K_c)^{-1}, \tag{6.9}$$

where G_1 is the sensitivity function, and G_3 is the complementary sensitivity function.

Then, the weighted mixed-sensitivity can be obtained.

$$\begin{bmatrix} W_1 e \\ W_2 M_B \\ W_3 RI \\ e \end{bmatrix} = \begin{bmatrix} W_1 & -W_1 G_s \\ 0 & W_2 \\ 0 & W_3 G_s \\ I & -G_s \end{bmatrix} \begin{bmatrix} r_{in} \\ M_B \end{bmatrix}. \tag{6.10}$$

Appling the minimum gain theorem, in case of mixed sensitivity problem, the objective is to find a rational function controller and to make the rollover prevention closed-loop system stable satisfying the following inequality.

$$\left\| \begin{matrix} W_1 G_1 \\ W_2 G_2 \\ W_3 G_3 \end{matrix} \right\|_\infty < 1. \tag{6.11}$$

The weight functions W_1, W_2, and W_3 are the tuning parameters and it typically requires some iterations to obtain weights which will yield a good controller. For a good robustness margin and a small tracking error, a good starting point is to choose them as:

$$W_1 = \frac{k_1 s + k_2}{k_3 s + 1} \tag{6.12}$$

$$W_2 = k_4 \tag{6.13}$$

$$W_3 = \frac{k_5 s + k_6}{k_7 s + 1}. \tag{6.14}$$

There are various methods available in the literature for selection of weights. In most of these design methods the weight functions are selected using trial and error method.

Then, H-infinity controller is synthesized by loop shaping technique. But there is a disadvantage in this type of synthesis that trial and error procedure may not end up in a stabilizing controller. So, the genetic algorithm is used to select the optimal weight functions and the flowchart of H-infinity controller programming with genetic algorithm is given as shown in Figure 6.7.

The parameters $k_1, k_2, k_3, k_4, k_5, k_6$, and k_7 are to be optimized with genetic algorithm, of which the variation range can be limited by the dynamic model of vehicle rollover. The objective function of the genetic optimization J is defined as the maximum absolute value of rollover index. And the fitness function S_H is defined as the reciprocal of objective function.

$$\begin{cases} J \overset{def}{=} \max(|RI|) \\ S_H \overset{def}{=} 1/J. \end{cases} \tag{6.15}$$

Also, to meet the stability criterion of H-infinity controller for vehicle rollover prevention, the constraint condition in inequality (6.16) should be satisfied:

$$\bar{\sigma}\left[W_1^{-1}(j\omega)\right] + \bar{\sigma}\left[W_3^{-1}(j\omega)\right] \geq 1. \tag{6.16}$$

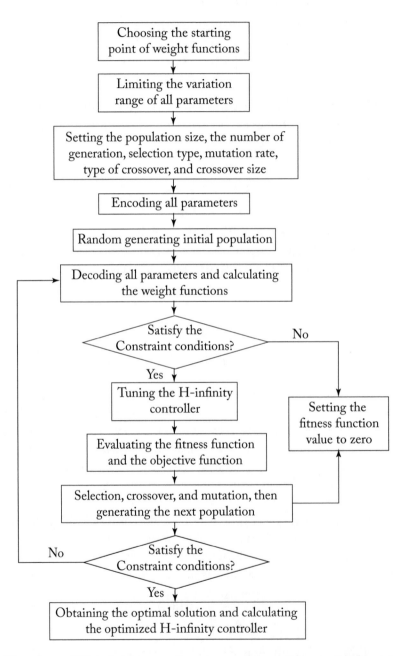

Figure 6.7: Flowchart of H-infinity controller programming with genetic algorithm.

Using the optimization algorithm shown in Figure 6.7, the optimized solution can be obtained and the results show that $k_1 = 0.5, k_2 = 2, k_3 = 1, k_4 = 1e-5, k_5 = 0.01, k_6 = 1e-5$, and $k_7 = 1$.

The optimized H-infinity controller can be derived as Equation (6.17) based on the parameters a large passenger vehicle.

$$K_c = \frac{\left\{ \begin{array}{l} 1.665e4s^{11} + 1.524e6s^{10} + 1.404e8s^9 + 5.811e9s^8 + 2.262e11s^7 + 3.79e12s^6 \\ +6.125e13s^5 + 5.565e14s^4 + 4.502e15s^3 + 2.214e16s^2 + 6.327e16s + 4.511e16 \end{array} \right\}}{\left\{ \begin{array}{l} s^{11} + 101.7s^{10} + 9172s^9 + 4.195e5s^8 + 1.569e7s^7 + 3.149e8s^6 + 3.965e9s^5 \\ +5.021e10s^4 + 2.842e11s^3 + 2.208e12s^2 + 3.754e12s + 1.784e12 \end{array} \right\}}.$$

(6.17)

Furthermore, the optimized solutions and controller varies with different value of parameters. The gains can be calculated offline for different parameters and using a look up table they are selected as the parameters are varied.

To minimize the effect of disturbance on the output, the sensitivity function and the complementary sensitivity function should be reduced. Also, the system must be robust enough to provide good performance and stability over the uncertainty. So, the following constraints should be met:

$$\left\{ \begin{array}{l} \bar{\sigma}\left[G_1(j\omega)\right] < \bar{\sigma}\left[W_1^{-1}(j\omega)\right] \\ \bar{\sigma}\left[G_3(j\omega)\right] < \bar{\sigma}\left[W_3^{-1}(j\omega)\right] \end{array} \right. .$$

(6.18)

The constraints are based on the singular values which are good measures of the system robustness. Figure 6.8 plots the singular value plot of the system with H-infinity control which shows the relationship of amplitude and frequency of the sensitivity function, the complementary sensitivity function, the performance weight function W_1, and the robustness weight function W_3. As shown in Figure 6.8, the amplitude of the sensitivity function is small in low frequencies. The singular value curve of the sensitivity function is lower than that of the performance weight function, and the singular value curve of the complementary sensitivity function is lower than that of the robustness weight function. So, the optimized H-infinity controller operates in a stable environment and provides good control for the vehicle rollover system.

Two typical driving conditions are used to simulate the untripped rollover stability of the vehicle with the controller, i.e., Fishhook and double-lane change maneuver.

Figure 6.9 compares the new rollover index of the vehicle in Fishhook case with different control strategies, including without control, traditional proportional integral derivative (PID) control method, and optimized H-infinity control method. The traditional PID controller is tuned by the critical proportion method and the gains are set K_P to 5000, K_I to 20, and K_D to 500. As shown in Figure 6.9, the absolute value of the rollover index of the vehicle is over 1 at 2.57 s without extra brake force, so the vehicle rolls over. While the rollover will be prevented by differential braking force at each wheel with the traditional PID control method or the optimized H-infinity control method. Furthermore, the maximum absolute value of the rollover index of

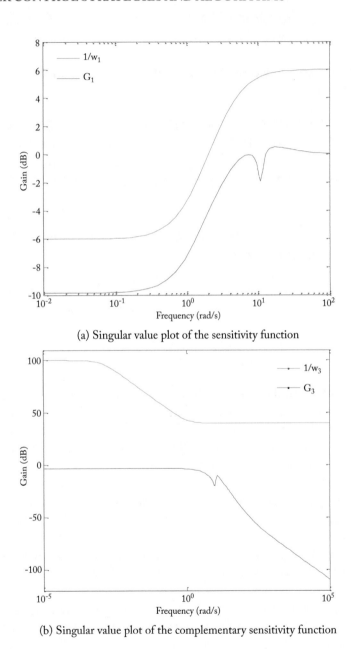

(a) Singular value plot of the sensitivity function

(b) Singular value plot of the complementary sensitivity function

Figure 6.8: Singular value plot of the system with H-infinity control.

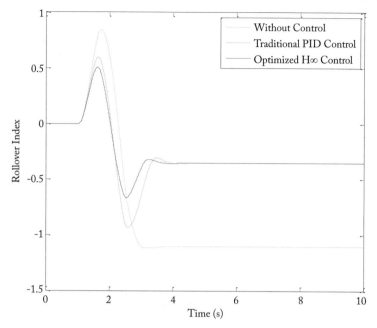

Figure 6.9: Rollover indices of the vehicle under the Fishhook case.

the vehicle with the optimized H-infinity control method is lower and varies more smoothly than that with the traditional PID control method.

Figure 6.10 show the results when the vehicle moves under the double-lane change maneuver. The similar conclusions can be drawn as under the Fishhook case. So, this rollover avoidance control system has a good robustness for different untripped driving situations.

To demonstrate the performance of the rollover avoidance control system when vehicle runs on a road with bumps, two rollover cases are selected, as follows.

Case I

In a tripped rollover situation, the vehicle rollover happens due to external road input, such as an unpredictable road bump under the right wheel when vehicle moves on a straight lane. The maximum height of the road bump is 0.15 m, and the vehicle speed is 100 km/h.

Case II

In this case a combined untripped and tripped rollover due to an unpredictable road bump under the right wheel while driving in a step steering is studied. The final value of the steering angle of the front wheel is $\delta = 2°$, the maximum height of the road bump is 0.15 m.

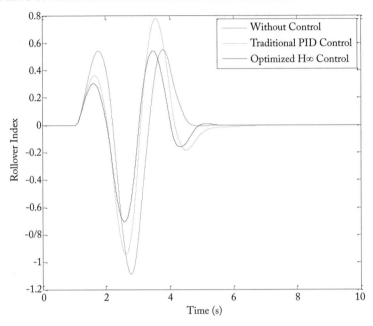

Figure 6.10: Rollover indices of the vehicle under the double-lane change maneuver.

As shown in Figure 6.11, the maximum absolute value of the rollover indices are more than 1 for the vehicle without a controller. So, the vehicle rolls over when it moves in a straight line or at cornering with an unpredictable bump without control. However, the rollover risk can be avoided by the differential braking force using a traditional PID control method and optimized H-infinity control method. Also, it can be found that the maximum value of the rollover index of vehicle with the optimized H-infinity control method is lower and varies more smoothly than that with the traditional PID control method in a tripped rollover situation. Therefore, the optimized H-infinity controller can obviously prevent vehicle rollover in a tripped rollover situation.

6.3 MODEL PREDICTION CONTROL METHOD

Model predictive control features good control effect, strong robustness, and low requirement for model accuracy and it can be used to control complex process effectively. So, model predictive control is designed to prevent vehicle rollover by many researchers [35, 64, 65].

Model prediction control consists of three modules: MPC controller, controlled object, and state solver. As shown in Figure 6.12, the MPC controller is responsible for combining the prediction model, objective function and constraint condition to obtain the optimization solution, getting the optimal control sequence $u^*(t)$ of the current moment, then sending the

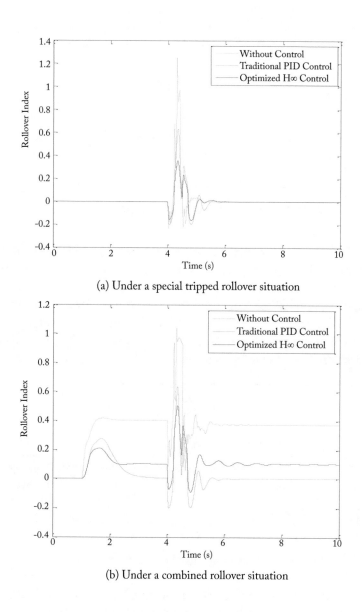

(a) Under a special tripped rollover situation

(b) Under a combined rollover situation

Figure 6.11: Comparison of the rollover indices under different situations.

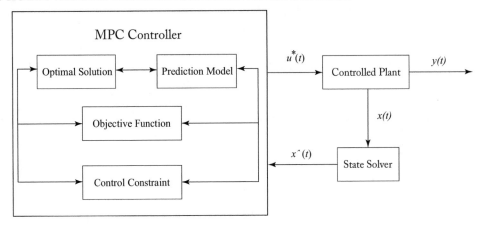

Figure 6.12: Functional block diagram of Model Predictive Control.

$u^*(t)$ into controlled object. After that, some state values $x(t)$ of current collection are sent to state solver. The state solver solves or estimates the state quantity $\hat{x}(t)$ that cannot be obtained directly from the sensor according to the current state value. After that, $\hat{x}(t)$ is sent into the MPC controller, make optimal solution again, and the future control sequence is obtained. The control process of MPC is formed through the calculation of reciprocating.

Based on MPC, the strategy of vehicle rollover control is analyzed based on the feedback of roll angle and yaw rate, as shown in Figure 6.13.

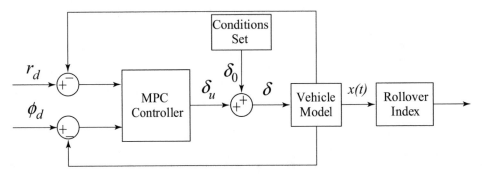

Figure 6.13: MPC rollover control strategy based on active steering.

The actual front-wheel steering angle δ is obtained by the accumulation of front-wheel steering angle δ_0 under setting conditions and control quantity of front-wheel steering angle δ_u output by MPC controller. And the controller calculates the control quantity of front-wheel steering angle δ_u based on the difference between the actual roll angle ϕ and desired roll angle ϕ_d, and the difference between the actual yaw rate r and desired yaw rate r_d. $x(t)$ is the vehicle states which is used to obtain the rollover index. The desired roll angle $\phi_d = 0$ and the desired yaw rate can be obtained from the linear 2-DOF model. That is:

$$r_d = \frac{u\delta}{L\left(1 + K_e u^2\right)},$$ (6.19)

where K_e is the gain of yaw rate.

MPC controller uses following objective function:

$$J = \sum_{j=1}^{P_h} \|y\left(k + j\right) - y_d\left(k + j\right)\|_Q^2 + \sum_{j=1}^{C_h} \|\Delta U\left(k + j\right)\|_R^2.$$ (6.20)

In this equation, P_h is prediction horizon, C_h is control horizon, Q is the output state weighting coefficient, R is the control weight coefficient, y represents output state, y_r represents output reference value, and ΔU is the control increment.

Moreover,

$$y\left(k + j\right) = \begin{bmatrix} \phi\left(k + j\right) \\ r\left(k + j\right) \end{bmatrix}, \qquad y_d\left(k + j\right) = \begin{bmatrix} 0 \\ r_d\left(k + j\right) \end{bmatrix}.$$ (6.21)

The vehicle model mentioned in Section 2.3 is used as predictive model and the state-space equation of predictive model is shown in Equation (6.22):

$$\dot{x} = Ax + B\delta_u,$$ (6.22)

where x is the rollover state quantities of vehicle. A and B are the state space equation matrix.

The model is discretized. According to the first order differential quotient method, the model can be discretized as discrete state-space equation as follows:

$$\begin{cases} x(k + 1) = A_k x(k) + B_k \delta_u(k) \\ y(k) = C_k x(k) \end{cases}$$ (6.23)

of which

$$\begin{cases} A_k = E + t_s A \\ B_k = t_s B. \end{cases}$$ (6.24)

In Equations (6.23) and (6.24), E is unit matrix which has the same dimension as A, C_k is the coefficient matrix in the output equation of state-space equation, t_s is discrete sampling time, and $t_s = 0.001$.

The objective function needs to calculate the state output of the future time:

$$\chi(k) = \begin{bmatrix} x(k) \\ \delta_u(k) \end{bmatrix}.$$ (6.25)

So, the state-space equation of prediction can be transformed into:

$$\begin{cases} \chi(k+1) = \hat{A}_k \chi(k) + \hat{B}_k \Delta \delta_u(k) \\ y(k) = \hat{C}_k \chi(k), \end{cases} \tag{6.26}$$

where

$$\hat{A}_k = \begin{bmatrix} A_k & B_k \\ 0_{1\times 8} & 1 \end{bmatrix}, \quad \hat{B}_k = \begin{bmatrix} B_k \\ 1 \end{bmatrix}, \quad \hat{C}_k = \begin{bmatrix} C_k & 0 \end{bmatrix}.$$

So, the output prediction equation is

$$Y(k) = \psi_k \chi(k) + \Theta_k \Delta \delta_u(k) \tag{6.27}$$

of which

$$Y(k) = \begin{bmatrix} y(k+1) \\ y(k+2) \\ \cdots \\ y(k+C_h) \\ \cdots \\ y(k+P_h) \end{bmatrix} \qquad \psi(k) = \begin{bmatrix} \hat{C}_k \hat{A}_k \\ \hat{C}_k \hat{A}_k^2 \\ \cdots \\ \hat{C}_k \hat{A}_k^{C_h} \\ \cdots \\ \hat{C}_k \hat{A}_k^{P_h} \end{bmatrix}$$

$$\Theta(k) = \begin{bmatrix} \hat{C}_k \hat{B}_k & 0 & 0 & 0 \\ \hat{C}_k \hat{A}_k \hat{B}_k & \hat{C}_k \hat{B}_k & 0 & 0 \\ \cdots & \cdots & \ddots & \cdots \\ \hat{C}_k \hat{A}_k^{C_h-1} \hat{B}_k & \hat{C}_k \hat{A}_k^{C_h-2} \hat{B}_k & \cdots & \hat{C}_k \hat{B}_k \\ \hat{C}_k \hat{A}_k^{C_h} \hat{B}_k & \hat{C}_k \hat{A}_k^{C_h-1} \hat{B}_k & \cdots & \hat{C}_k \hat{A}_k \hat{B}_k \\ \cdots & \cdots & \ddots & \cdots \\ \hat{C}_k \hat{A}_k^{P_h-1} \hat{B}_k & \hat{C}_k \hat{A}_k^{P_h-2} \hat{B}_k & \cdots & \hat{C}_k \hat{A}_k^{P_h-C_h-1} \hat{B}_k \end{bmatrix}$$

$$\Delta U(k) = \begin{bmatrix} \Delta \delta_u(k) \\ \Delta \delta_u(k+1) \\ \cdots \\ \Delta \delta_u(k+C_h) \end{bmatrix}.$$

Because the vehicle model is changing in real time, in order to guarantee the objective function can obtain the feasible solution at every moment. So, add the relaxation factor to the optimization goal:

$$J(\chi(k), \delta_u(k-1), \Delta U(k)) = \sum_{j=1}^{P_h} \|y(k+j) - y_d(k+j)\|_Q^2 + \sum_{j=1}^{C_h} \|\Delta U(k+j)\|_R^2 + \rho \varepsilon_1^2, \tag{6.28}$$

where ρ is weight coefficient and ε_1 is relaxation factor.

Because the steering angle of vehicle is limited, physical constraint must be considered when designing the controller. So controlled quantity of steering angle should be constraint as follows:

$$\delta_{u_{\min}}(k+j) \leq \delta_u(k+j) \leq \delta_{u_{\max}}(k+j), \quad j = 0, 1, \dots, C_h - 1. \tag{6.29}$$

Because

$$\delta_u(k+j) = \delta_u(k+j-1) + \Delta\delta_u(k+j), \tag{6.30}$$

assume

$$U_k = 1_{C_h} \otimes \delta_u(k-1) \tag{6.31}$$

$$A_{C_h \times C_h} = \begin{bmatrix} 1 & 0 & \cdots & 0 \\ 1 & 1 & 0 & \vdots \\ \vdots & 1 & \ddots & 0 \\ 1 & \cdots & 1 & 1 \end{bmatrix} \otimes E_1. \tag{6.32}$$

In Equation (6.32), 1_{C_h} is column vector which has C_h rows, E_1 is unit matrix and the dimension of E_1 is 1, $\otimes$ represents Kronecker product, and $\delta_u(k-1)$ represents the actual control output at the last moment.

According to Equations (6.30) and (6.31), Equation (6.29) can be transformed as:

$$U_{\min} \leq A_{C_h \times C_h} \Delta U_k + U_k \leq U_{\max}, \tag{6.33}$$

where $U_{\min}$ and $U_{\max}$ are the collection of maximal and minimum values of control in time domain.

The complete objective function can be obtained by substituting Equation (6.27) into the objective function (6.28). Then, the objective function is reduced to a standard quadratic form:

$$J(\chi(k), \delta_u(k-1), \Delta U(k)) = \left[\Delta U(k)^T, \varepsilon\right]^T H_k \left[\Delta U(k)^T, \varepsilon\right] + G_t \left[\Delta U(k)^T, \varepsilon\right] + P_t \tag{6.34}$$

where

$$H_k = \begin{bmatrix} \Theta_k^T Q \Theta_k + R & 0 \\ 0_Q & \rho \end{bmatrix}, \quad G_k = \begin{bmatrix} 2E(k)Q\Theta_k & 0_Q \end{bmatrix}, \quad P_k = E(k)^T Q E(k),$$

where 0_Q is the null matrix that has the same dimension as Q matrix, $E(k)$ is tracking error in prediction horizon, and it can be showed as:

$$E(k) = \psi(k)\chi(k) - Y_{ref}(k), Y_{ref}(k)$$
$$= \left[y_{ref}(k+1), y_{ref}(k+2), \dots, y_{ref}(k+P_h)\right]^T, \tag{6.35}$$

where $y_{ref}(k)$ is reference output.

Binding constraint condition (6.29), a control increment sequence in control horizon can be obtained by solving Equation (6.28):

$$\Delta U_k^* = \left[\begin{array}{cccc} \Delta (\delta_u)_k^* & \Delta (\delta_u)_{k+1}^* & \cdots & \Delta (\delta_u)_{k+C_h-1}^* \end{array} \right]^T. \tag{6.36}$$

The first element in the control increment sequence is used as the actual control increment for the control system:

$$\delta_u(k) = \delta_u(k-1) + \Delta (\delta_u)_k^*. \tag{6.37}$$

In the following control cycle, the continuous cycle of the above solution can realize the rollover control.

The control output quantity as $[-24, 24]$. After debugging, the parameters of MPC controller are evaluated, prediction horizon $P_h = 35$, control horizon $C_h = 5$, weight coefficient $\rho = 10$, relaxation factor $\varepsilon = 10$, output state coefficient $Q = [Q_1 \ 0_{8 \times 27}]$, control weight coefficient $R = r_1 [1 \ 0_{1 \times 4}]$, where

$$Q_1 = \begin{bmatrix} Q_q & \cdots & 0_{2\times 2} \\ \vdots & \ddots & \vdots \\ 0_{2\times 2} & \cdots & Q_q \end{bmatrix}, \qquad Q_q = \begin{bmatrix} 3.33 & 0 \\ 0 & 5 \end{bmatrix}, \qquad r_1 = 0.3.$$

Under the Fishhook condition, there are two phases that are prone to roll over. Therefore, it is very necessary to analyze the control effect of multi-objective MPC control under this condition. The initial velocity is 85 km/h and the maximum front-wheel angle input is six degrees. The comparison of control effect between multi-objective MPC control and PID control is shown in Figure 6.14.

Figures 6.14a–d, respectively, represent the comparison of *mRI*, yaw rate, roll angle, and steering angle under the Fishhook maneuver. As we can see in Figure 6.14a, the three-axis bus rolls over at 4 s without control. Both the MPC control and PID control can prevent the bus from rollover. Obviously, the MPC control can make the *mRI* smaller. In Figure 6.14b, it shows the yaw rate with MPC control is more consistent with desired value. It suggests that the MPC control can follow the steering intention of the driver while ensuring the control effect. Figure 6.14c shows the roll angle is almost four degrees without control which is very dangerous. The MPC control limits it to two degrees which is smaller than PID control. To sum up, it can be concluded that MPC control can effectively prevent vehicle rollover.

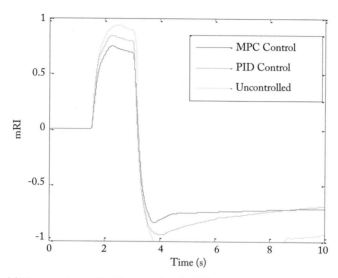

(a) Comparison of rollover index (mRI) under the Fishhook condition

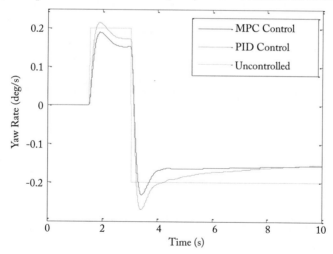

(b) Comparison of yaw rate under the Fishhook condition

Figure 6.14: Control effect comparison between MPC and PID control in the Fishhook condition. (*Continues.*)

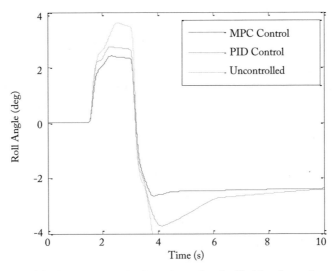

(c) Comparison of roll angle under the Fishhook condition

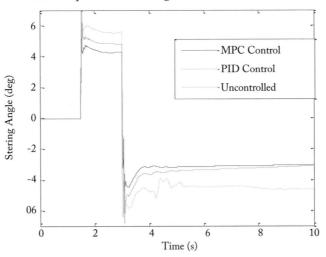

(d) Comparison of steering angle under the Fishhook condition

Figure 6.14: (*Continued.*) Control effect comparison between MPC and PID control in the Fishhook condition.

6.4   LINEAR QUADRATIC REGULATOR CONTROL METHOD

LQR is an algorithm for optimal control by taking the integral of quadratic function of state variable as the performance index of a linear system. It greatly simplifies the computation of real-time control. So, it also widely used in the development of rollover avoidance control system [44, 66–68]. Yim et al. proposed a linear quadratic regulator for vehicle rollover prevention as follows, which combined the Electronic Stability Control (ESC) and Active Roll Control System (ARCS) [69].

First, 3-DOF model is established, consisting of a 2-DOF bicycle model to describe yaw motion and lateral motion, and a 1-DOF roll model to describe the roll motion.

The LQR cost function for rollover prevention is defined as:

$$J = \int_0^\infty \left(\rho_1 e_\gamma^2 + \rho_2 a_y^2 + \rho_3 \phi^2 + \rho_4 \dot{\phi}^2 + \rho_5 M_B^2 + \rho_6 M_\phi^2 \right) dt. \tag{6.38}$$

In Equation (6.38), the weights ρ_i are set by the relation $\rho_i = 1/\eta_i^2$ from Bryson's rule, where η_i represents the maximum allowable value of each term, a_y is the lateral acceleration of vehicle, ϕ is the roll angle, and $\dot{\phi}$ is roll rate. M_B is the controlled yaw moment generated by differential braking and M_ϕ is the roll moment induced by active anti-roll bar. e_γ is the yaw rate error and is defined as the difference between the actual yaw rate r and the reference yaw rate r_d, according to

$$e_\gamma = r - r_d. \tag{6.39}$$

The reference yaw rate γ_d generated from the driver's steering input δ is modeled with a first-order system as

$$r_d = \frac{K_e}{\tau s + 1} \delta, \tag{6.40}$$

where τ is the time constant and K_e is the steady-state yaw rate gain determined by the speed of the vehicle.

The yaw dynamics are modeled separately without the roll dynamics. For the 3-DOF vehicle model, the state x, the control input u, and the disturbance w are defined as

$$\begin{cases} x = \begin{bmatrix} v & r & r_d & \phi & \dot{\phi} \end{bmatrix}^T \\ u = \begin{bmatrix} M_B & M_\phi \end{bmatrix} \\ w = \delta, \end{cases} \tag{6.41}$$

where v is lateral speed.

Based on the 3-DOF vehicle rollover model, the state-space equation of the vehicle model is obtained as

$$\dot{x} = Ax + B_1 w + B_2 u, \tag{6.42}$$

where A, B_1, and B_2 are the coefficient matrix of the state space equation of the 3-DOF vehicle rollover model. These matrixes can be derived by vehicle rollover model.

To avoid rollover in cornering situations, the roll angle and the roll rate should be reduced under the assumption that the lateral acceleration is controllable. If the weights on the roll angle and the roll rate are set to higher values for rollover prevention, the yaw rate error increases owing to the lateral load transfer caused by the ARCS, and this can make a vehicle lose maneuverability or lateral stability because of the rear-sway phenomenon. On the other hand, if the weight on the yaw rate error is set to a higher value for maneuverability or lateral stability, the roll angle and the roll rate cannot be reduced effectively. These effects are complementary to each other. In other words, the ARCS can reduce the roll angle and roll rate which ESC cannot do, and ESC can reduce the yaw rate error which the ARCS cannot do. For this reason, the weights on the roll angle, the roll rate and the yaw rate error, i.e., ρ_1, ρ_3, and ρ_4 in Equation (6.38), should be set to higher values in the LQ objective function. The values of η_i for the weights in the LQR cost function are given in Table 6.1.

Table 6.1: Weights in the LQ cost function

η_1	η_2	η_3	η_4	η_5	η_6
0.08 rad/s	10 m/s^2	1°	3 deg/s	5000 N•m	2000 N•m

The LQR cost function in Equation (6.36) can be rewritten as

$$J = \int_0^\infty (Cx + Du)^T (Cx + Du)dt$$
$$= \int_0^\infty \left(x^T Q x + u^T N^T x + x^T N^T u + u^T R u\right) dt,$$

(6.43)

where $Q = C^T C$, $N = C^T D$, $R = D^T D$

$$C = \begin{bmatrix} \sqrt{\rho_1}a_{11} & \sqrt{\rho_1}(a_{12} + v_x) & \sqrt{\rho_1}a_{13} & \sqrt{\rho_1}a_{14} & \sqrt{\rho_1}a_{15} \\ 0 & \sqrt{\rho_2} & 0 & 0 & -\sqrt{\rho_2} \\ 0 & 0 & \sqrt{\rho_3} & 0 & 0 \\ 0 & 0 & 0 & \sqrt{\rho_4} & 0 \\ 0 & 0 & 0 & 0 & 0 \\ 0 & 0 & 0 & 0 & 0 \end{bmatrix} \quad D = \begin{bmatrix} 0 & 0 \\ 0 & 0 \\ 0 & 0 \\ 0 & 0 \\ \sqrt{\rho_5} & 0 \\ 0 & \sqrt{\rho_6} \end{bmatrix}.$$

In Equation (6.43), a_{ij} is the jth element of the ith row of the matrix A. In the LQR, the full-state feedback control $u = Kx$ is used. The controller gain K is easily obtained by solving the Riccati equation. Then the LQR controller is designed.

The controllers for the yaw motion and the roll motion can be designed separately if the ARCS can reduce the roll motion. To confirm this fact, a structured controller and a separated controller are designed based on LQR controller.

In a structured controller, the control yaw moment is calculated only from the yaw rate error, and the control roll moment is calculated only from the roll angle and the roll rate. The controller structure is given by

$$K_s = \begin{bmatrix} 0 & k_1 & -k_1 & 0 & 0 \\ 0 & 0 & 0 & k_2 & k_3 \end{bmatrix}. \tag{6.44}$$

In a structured LQR controller, it is necessary to find K_s which minimizes the LQR cost function J. With the notation in Equation (6.43), the problem of finding K_s is formulated as the optimization problem.

$$\min_K \quad J = trace(P)$$

$$s.t. \qquad \begin{aligned} (A + B_2 K_s)^T P + P (A + B_2 K_s) + Q \\ + K_s^T N^T + K_s N + K_s^T R K_s. \end{aligned} \tag{6.45}$$

For an arbitrary K_s, the LQR cost function J can be easily computed by solving the Lyapunov Equation (6.45). The evolutionary strategy with covariance matrix adaptation (CMA-ES)12 is used to find the optimal K_s. The structured controller given by Equation (6.44) is designed for the single objective function (6.38). In this situation, the controller gains for the yaw moment and the roll moment have an effect on each other. This means that the structured controller does not strictly separate the yaw motion and the roll motion. Hence, it is necessary to design separately a controller for yaw motion and a controller for roll motion with different objective functions. Therefore, two LQR controllers are designed separately for yaw motion control and roll motion control. For this purpose, the LQR objective function given by Equation (6.38) is separated into J_{s1} and J_{s2} according to the yaw motion and the roll motion as given by

$$J_{s1} = \int_0^\infty \left(\rho_1 e_\gamma^2 + \rho_2 a_y^2 + \rho_5 M_B^2 \right) dt$$

$$J_{s2} = \int_0^\infty \left(\rho_3 \phi^2 + \rho_4 \dot{\phi}^2 + \rho_6 M_\phi^2 \right) dt. \tag{6.46}$$

With these objective functions and the state-space Equations (6.42), (6.47), and (6.48) represent the state-space equations of the yaw motion and roll motion, respectively, which are used to design separated controllers:

$$\begin{cases} x_\gamma = \begin{bmatrix} v & r & r_d \end{bmatrix} \\ u_\gamma = M_B \\ w_\gamma = \delta \\ \dot{x}_\gamma = A_\gamma x_\gamma + B_{1\gamma} w_\gamma + B_{2\gamma} u_\gamma \end{cases} \tag{6.47}$$

$$\begin{cases} x_\gamma = \begin{bmatrix} \phi & \dot{\phi} \end{bmatrix}^T \\ u_\gamma = M_\phi \\ w_\phi = a_y \\ \dot{x}_\phi = A_\phi x_\phi + B_{1\phi} w_\phi + B_{2\phi} u_\phi. \end{cases} \quad (6.48)$$

Accordingly, the structured controller gain matrix (6.44) is also separated as

$$\begin{aligned} K_{s1} &= \begin{bmatrix} 0 & k_1 & -k_1 \end{bmatrix} \\ K_{s2} &= \begin{bmatrix} k_2 & k_3 \end{bmatrix}. \end{aligned} \quad (6.49)$$

With these state-space equations, the objective functions and the controller structures, K_{s1} and K_{s2} are designed to minimize J_{s1} and J_{s2}, respectively. K_{s1} can be regarded as the proportional controller gain matrix for yaw motion control, and K_{s2} can be regarded as the proportional—derivative controller gain matrix for roll motion control. To find K_{s1} and K_{s2}, the CMA-ES is also used.

Figure 6.15 shows the Bode plots drawn on the basis of the designed LQ controllers. In these plots, the input is the steering angle, and the outputs are the roll angle, the roll rate, the lateral acceleration, and the yaw rate error. Since the frequency of the Fishhook maneuvre with a maximum steering angle of 221° is near 0.5 Hz, the frequency responses near 0.5 Hz should be focused on.

As shown in Figure 6.15, the structured LQR controller and the separated LQR controllers show almost the same performance. This is caused by the fact that the yaw motion and the roll motion of the controlled vehicle were not related to each other. This result means that the controllers for the yaw motion and the roll motion can be separately designed according to their own objectives and that several approaches can be applied to design the controllers for the yaw motion and the roll motion.

6.5   SLIDING MODE CONTROL METHOD

Sliding mode control (SMC) is a nonlinear control technique featuring remarkable properties of accuracy, robustness, and easy tuning and implementation. So, sliding mode control has attracted the attention of more and more scholars. Imine et al. proposed an estimator based on the high-order sliding mode observer is developed to estimate the vehicle dynamics, such as lateral acceleration limit and center height of gravity [70]. An antiroll controller is designed by Chuwith smooth sliding mode control technique for vehicles in which an active antiroll suspension is installed [33].

A rollover avoidance controller with sliding mode technique to control the steering angle of vehicle is designed. And two sliding mode control are developed first, and then optimized them into one controller which has a better performance with the theory of fuzzy control. Taking

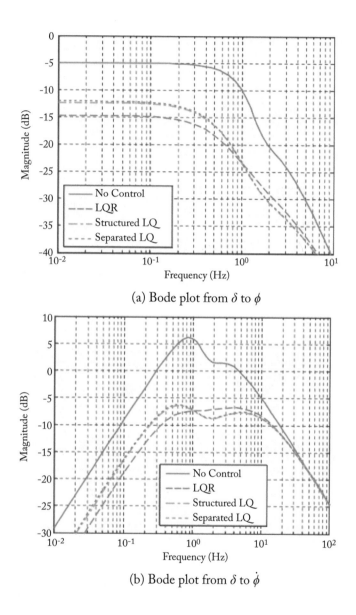

(a) Bode plot from δ to ϕ

(b) Bode plot from δ to $\dot{\phi}$

Figure 6.15: Bode plots from the steering input to each output. (*Continues.*)

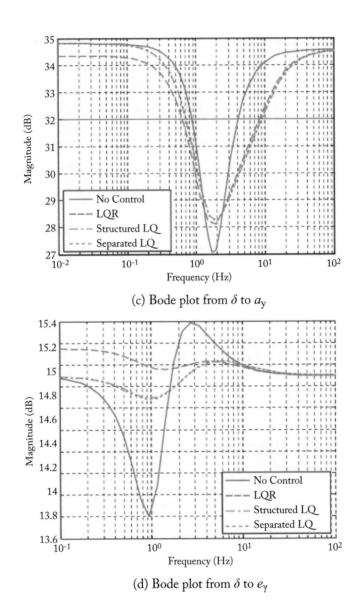

(c) Bode plot from δ to a_y

(d) Bode plot from δ to e_γ

Figure 6.15: (*Continued.*) Bode plots from the steering input to each output.

into account the true steering actuator may not be able to complete the output response control, so set the control output of the rotation angle is $\pm 2.3°$, and the maximum change rate of control output is $\pm 10°/\text{s}$ (see Figure 6.16).

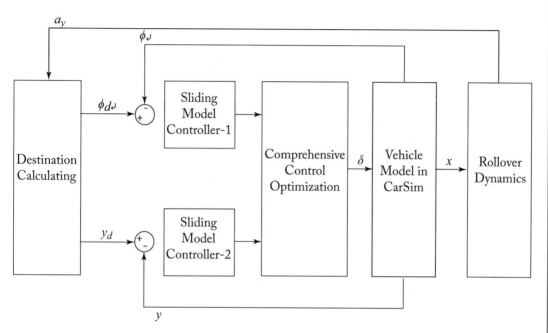

Figure 6.16: The rollover prevention controller diagram.

1. Sliding mode control in consideration of roll rate.

   Rewrite the vehicle rollover model in state form as follows.

   Setting the state vector as $x = \begin{bmatrix} v & r & \phi & \dot{\phi} & \phi_u \end{bmatrix}$, the state-space equation can be obtained based on the vehicle dynamics model:

   $$\dot{x} = Ax + B\delta. \tag{6.50}$$

   And discretize the system with the method of approximate discretization.

   Consider the sliding mode surface as:

   $$S(x) = \dot{e} + c_0 e, \qquad e = \phi_d - \phi, \tag{6.51}$$

   where ϕ_d is calculated from a saturation function:

   $$\begin{cases} \phi_d = 0.8\phi & |a_y| > 0.35 \\ \phi_d = \phi & |a_y| \le 0.35. \end{cases} \tag{6.52}$$

The control amount u_c is obtained from the related sliding mode reaching rule and the equations of deriving variable $\dot{S}(x)$.

$$U_c = flag\left(-U\,\text{sgn}(s(x))\right),\tag{6.53}$$

where

$$\begin{cases} flag = 1 & |a_y| > 0.35 \\ flag = 0 & |a_y| \leq 0.35. \end{cases}\tag{6.54}$$

So:

$$u_c = T_s(CB)^{-1}\left[-C\left(I + T_s A\right)(x - x_d)\right] + w_1 e - w_2 T_s(CB)^{-1}\text{sgn}(e),\tag{6.55}$$

where $w_1 = 1$, $C = w_2\,[1\ 1\ 1\ 1]$.

2. Sliding mode control in consideration of lateral displacement.

The lateral and yaw accelerations are computed as follows:

$$\begin{aligned} \ddot{y} &= a_1\dot{y} + a_2\dot{r} + a_3\delta \\ \ddot{r} &= b_1\dot{r} + b_2\delta + b_3\beta. \end{aligned}\tag{6.56}$$

The chose sliding mode surface is as follows:

$$S = \dot{y}_l + \eta y_l,\tag{6.57}$$

where

$$\dot{y}_l = \dot{y} - \dot{y}_d\tag{6.58}$$
$$y_l = y - y_d,\tag{6.59}$$

and $\dot{y}_d$, y_d are the first and double integration of desired lateral acceleration $a_{y\,\text{lim}}$, respectively.

Under the constant reaching rule:

$$\dot{S} = -\varepsilon\text{sgn}S \qquad \varepsilon > 0.\tag{6.60}$$

Assume that $x = \begin{bmatrix} \dot{y} \\ \dot{r} \end{bmatrix}$, and regard β as a disturbance, so Equation (6.50) can be recast as:

$$\dot{x} = \begin{bmatrix} a_1 & a_2 \\ 0 & b_1 \end{bmatrix} x + \begin{bmatrix} a_3 \\ b_2 \end{bmatrix}\delta.\tag{6.61}$$

Then

$$(a_1 + \eta)\,\dot{y} + a_2\dot{r} + a_3\delta = \ddot{y}_d + \eta\dot{y}_d - \varepsilon\text{sgn}(\varepsilon).\tag{6.62}$$

So

$$u_d = \delta = \frac{\left(\ddot{y}_d + \eta \dot{y}_d - \varepsilon \mathrm{sgn}(\varepsilon) - (a_1 + \eta)\, \dot{y} + a_2 \dot{r} \right)}{a_3}. \tag{6.63}$$

And then, a fuzzy control is used to optimized these two sliding mode controllers.

J-turn manueuver is used to verify the effectiveness of the sliding mode controllers. Vehicle speed in the simulations is set to 80 km/h and the default front-wheel steering angle in J-turn manueuver is set to 4.8°/s, which induces rollover easily. The dynamic responses of uncontrolled, first, and second kind of sliding mode control and the optimized controllers are shown in Figure 6.17.

Figure 6.17a shows that the trace under the first kind of control is far away from the desired trace and the trace under the second kind of control is almost consistent with the desired trace. The trace under the optimized control is between the traces under the first and second kind of control. The responses of rollover dynamic index LTR in Figure 6.17b shows that all three control methods can prevent the vehicle rollover. However, the value of LTR in the second kind of control is nearly reaching 0.9, which means the vehicle is unstable. Instead, the value of LTR can be reduced to about 0.5 in 3 s under the first kind of control. And the value of LTR can be reduced under 0.8 through the optimized control method. From these two figures, it can be concluded that the first kind of control method can prevent the vehicle rollover effectively but cannot track the desired trace. The second kind of control method have good track tracking effect, but the response of LTR shows that this method is under the risk of rollover. The optimized control does combine the advantages of the first two controls. The tracking error is reduced more than 30 m, and the tracking error is stabilized within 10 m. The value of LTR reaches about 0.6, and rises slowly to about 0.78, so as to track the desired trace.

Figure 6.17c shows the front-wheel steering angle under the three-control method comparing the steering angle without control. The steering angle of the first kind of control varies greatly, while the shimmy appears in the second kind of control. Though this two-control method can reach the designed purpose separately, these situations can cause occupant discomfort and adverse factors to steering system. The optimized control reduces the front-wheel angle variation, eliminate the shimmy, avoid the disadvantages of the first two control methods.

Figures 6.17d–f show the dynamic responses of roll state which includes roll angle and roll rate and yaw rate. The response of roll rate is similar with the response of the LTR. The first kind of control method shows its advantages, and the performance of optimized control method is between the first and second kind of control method. The response of yaw rate of the first kind of control is little, but it cannot support the need of trace tracking, as the steering angle is less than the default value in Figure 6.17f. The response of yaw rate of the second kind of control changes greatly, which will bring discomfort to the occupant. The value and variation of the response of yaw rate of the optimized control are between the first two kinds of control.

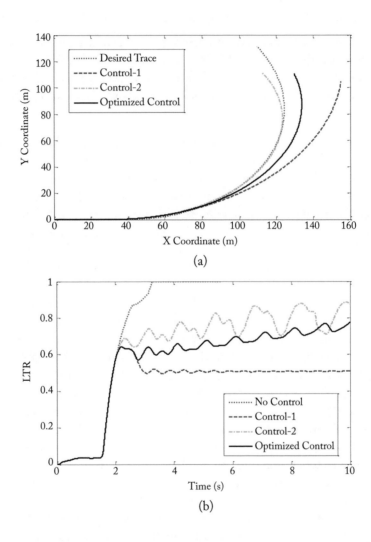

Figure 6.17: The control performance verification under J-turn maneuver. (*Continues.*)

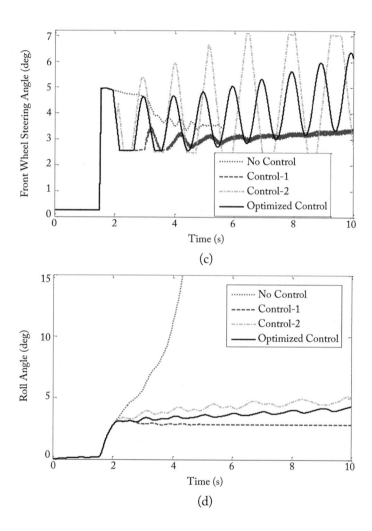

(c)

(d)

Figure 6.17: (*Continued.*) The control performance verification under J-turn maneuver. (*Continues.*)

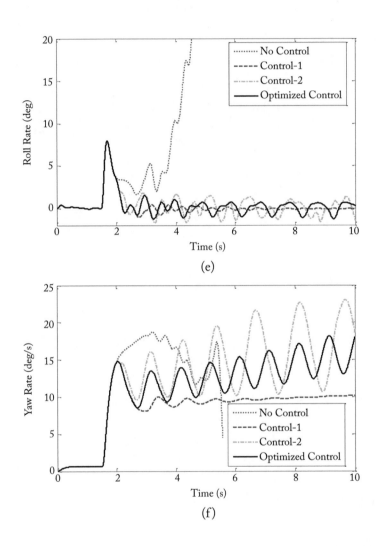

(e)

(f)

Figure 6.17: (*Continued.*) The control performance verification under J-turn maneuver.

6.6   SUMMARY

To sum up, there are many kinds of rollover control strategies and algorithms used to prevent vehicle rollover such as PID control, LQR control, H-infinity control controller, sliding mode control, and MPC control. To achieve better control effect, it is better to optimize the controller for different vehicles or employ the combination of one or more controller to complement each other.

CHAPTER 7

Conclusions

In this book, some different dynamic models are used to describe the vehicle rollover under both untripped and special tripped situations. From the vehicle dynamics theory, rollover indices are deduced and the dynamic stabilities of vehicle rollover are analyzed. In addition, some active control strategies are discussed to improve the anti rollover performance of the vehicle.

In Chapter 2, in order to more accurately describe the vehicle rollover motion, some vehicle rollover dynamic models are introduced, including: roll plane model, raw-roll model, lateral-yaw-roll model, yaw-roll-vertical model, multi-freedom model, and multi-body dynamic model.

In Chapter 3, in order to predict the vehicle rollover state, different rollover indexes of untripped rollover are discussed, including Static Stability Factor, Dynamics Stability Factor, Lateral Load Transfer Ratio, Time-to-Rollover, and Prediction Rollover Warning.

In Chapter 4, the model and index of tripped vehicle rollover are presented in some special cases such as uneven roads and banked roads. Also, the dynamics of tripped vehicle rollover are simulated. And the energy method is introduced to predict tripped vehicle rollover using the vehicle roll energy and the rollover potential energy.

In Chapter 5, the widely used rollover avoidance control methods are summarized and analyzed, including anti-roll bar system, active suspension system, active steering system, active braking system, and integrated chassis system.

In Chapter 6, many kinds of rollover control strategies and algorithms used to prevent vehicle rollover are disused, such as PID control, LQR control, H-infinity control controller, sliding mode control, and MPC control. Also, the effectiveness of different rollover control algorithms are illustrated.

APPENDIX A

Notation

Symbol	Parameter	Symbol	Parameter
a_y	Lateral acceleration of vehicle	F_f	Lateral force of a front tire
d	Distance from the CG to front axle	F_r	Lateral force of a rear tire
F_y	Lateral tire force in tire y-axis (of wheel plane)	F_L	Vertical load on left tires
F_z	Vertical force on tire	F_R	Vertical load on right tires
$H_{CG,s}$	COG height of sprung mass respect to grand	h	Height of the roll center, measured upward from the road
H_{RC}	Height of roll axis respect to grand	I_x	Roll moment of inertia of the sprung mass, measured about the roll axis
k_ϕ	Roll stiffness of suspension	I_z	Yaw moment of inertia of the total mass, measured about the z axis
M_{RC}	Roll moment, measured about roll center	k_f	Cornering stiffness coefficient of a front tire
m_s	Sprung mass	k_r	Cornering stiffness coefficient of a rear tire
ϕ	Vehicle roll angle	L	Wheelbase of vehicle
ϕ_B	Road bank angle	m	Total mass of vehicle
δ	Steering wheel angle	r	Yaw rate of the sprung mass
a	Longitudinal distance to the front axle	g	Gravitational acceleration
b	Longitudinal distance to the rear axle	T_w	Track width of vehicle
C_f	Cornering stiffness values for the front tires	U	Forward speed of vehicle
C_r	Cornering stiffness values for the rear tires	v	Lateral speed of vehicle

$\dot{\phi}$	Roll rate of vehicle	c_{s2}	Equivalent damping coefficient of the right suspension
$\ddot{\phi}$	Roll acceleration of vehicle	z_{s1}	Vertical displacement of sprung mass on the left
k_{t1}	Vertical stiffness of the left tires	z_{s2}	Vertical displacement of sprung mass on the right
k_{t2}	Vertical stiffness of the right tires	β_f	Slip angle of the front tires
z_c	Vertical displacement of sprung mass	β_r	Slip angle of the rear tires
z_{u1}	Vertical displacement of the left unsprung mass	ϕ_s	Roll angle of the sprung mass
z_{u2}	Vertical displacement of the right unsprung mass	ϕ_u	Roll angle of the unsprung mass
z_{r1}	Road input of the left tires	H	Height of center of mass, measures upward from the road
z_{r2}	Road input of the right tires	F_{zi}	Vertical force of the other side tires
F_{s1}	Dynamic forces of the left suspension	θ_{cr}	Vehicle body critical title angle
F_{s2}	Dynamic forces of the right suspension	μ	Coefficient of road adhesion
M_B	Anti-yaw torque	δ_d	Driver's steering-wheel angle
k_{s1}	Vertical stiffness of the left suspension	τ_{sw}	Steering first-order time constant
k_{s2}	Vertical stiffness of the right suspension	SR	Steering ratio
c_{s1}	Equivalent damping coefficient of the left suspension	c_ϕ	Equivalent roll damping coefficient of suspension
a_{ylim}	Desired lateral acceleration	r_d	Desired yaw rate
F_{b2}	Brake force of front-outer wheel	$x(t)$	State quantities of vehicle
F_{b1}	Brake force of front-inner wheel	K_e	Gain of yaw rate
C_b	Flow coefficient of solenoid valve	P_h	Prediction horizon
A_b	Area of orifice throttle	C_h	Control horizon
K_b	Bulk modulus of the brake oil	Q	Output state coefficient
V_b	Volume of the wheel brake cylinder	R	Control weight coefficient

n	Index of solenoid valve	$y(k)$	Output state
p_w	Pressure of wheel brake cylinder	$y_r(k)$	Output reference value
q	Average flow rate	ΔU	Control increment
γ	Adiabatic index	A, B	Sate space equation matrix.
p_0	Initial pressure of the gas chamber of accumulator	C_k	Coefficient matrix in the output equation of state-space equation
V_0	Volume of the gas chamber of accumulator	t_s	Discrete sampling time
δ_u	Control quantity of front wheel steering angle	ρ	Weight coefficient
δ_0	Accumulation of front wheel steering angle	ε_1	Relaxation factor
ϕ_d	Desired roll angle	M_ϕ	Roll moment induced by active anti-roll bar
e_γ	Yaw rate error	y_d	Double integration of desired lateral acceleration
τ	Time constant	$\dot{y}_d$	Desired lateral displacement
y	Lateral displacement	u_c	Output of the sliding mode control in consideration of roll rate
$\dot{y}$	First integration of lateral acceleration	u_d	Output of the sliding mode control in consideration of lateral displacement.
m_{sf}	Sprung mass of front axle	m_{sr}	Sprung mass of rear axle
m_{uf}	Unsprung mass of front axle	m_{ur}	Unsprung mass of rear axle
b_1	Longitudinal distance from the CG to the middle axle	c_1	Longitudinal distance from the CG to the rear axle
h_f	Height between the center of front sprung mass and the roll center	h_r	Height between the center of rear sprung mass and the roll center
h_{uf}	Height of the center of the front unsprung mass, measured upward from the road	h_{ur}	Height of the center of the rear unsprung mass, measured upward from the road
h_c	Height of the roll center, measured upward from the road	h_{cf}	Height of the front roll center, measured upward from the road

h_{cr}	Height of the rear roll center, measured upward from the road	I_{Xf}	Roll inertia of the front sprung mass
I_{Xr}	Roll inertia of the front sprung mass	φ_{sf}	Roll angle of the front sprung mass
φ_{sr}	Roll angle of the rear sprung mass	φ_{uf}	Roll angle of the front unsprung mass
φ_{ur}	Roll angle of the rear unsprung mass	F_{Yr}	Lateral force of the tires at the virtual axle
M_r	Yaw moment caused by the virtual rear axle	k_f	Equivalent roll stiffness coefficient of the front suspension
k_r	Equivalent roll stiffness coefficient of the rear suspension	k_{uf}	Equivalent roll stiffness coefficient of the front unsprung mass
k_{uf}	Equivalent roll stiffness coefficient of the rear unsprung mass	l_f	Equivalent roll damping coefficient of the front suspension
l_r	Equivalent roll damping coefficient of the rear suspension	k_b	Torsion stiffness coefficient of vehicle frame
β_m	Slip angle of the middle tires	k_m	Cornering stiffness coefficient of a middle tire
m_u	Unsprung mass	l_e	Equivalent wheelbase of the triaxle bus

Bibliography

[1] National highway traffic safety administration, traffic safety facts 2012: A compilation of motor vehicle crash data from the fatality analysis reporting system and the general estimates system. pp. 77–85, U.S. Department of Transportation, Washington, DC, 2014. 1

[2] H. Huang, R. Yedavalli, and D. Guenther. Active roll control for rollover prevention of heavy articulated vehicles with multiple-rollover-index minimization. *Vehicle System Dynamics*, 50(3):471–493, 2012. DOI: 10.1115/dscc2010-4278. 1, 49

[3] G. Mattos, R. Grzebieta, M. Bambach, et al. Validation of a dynamic rollover test device. *International Journal of Crashworthiness*, 18(3):207–214, 2013. DOI: 10.1080/13588265.2013.772766. 1

[4] J. Gertsch and T. Shim. Interpretation of roll plane stability models. *International Journal of Vehicle Design*, 46(1):72–77, 2008. DOI: 10.1504/ijvd.2008.017070. 3, 21, 44

[5] G. Yu, H. Li, P. Wang, et al. Real-time bus rollover prediction algorithm with road bank angle estimation. *Chaos Solitons and Fractals the Interdisciplinary Journal of Nonlinear Science and Nonequilibrium and Complex Phenomena*, 89:270–283, 2016. DOI: 10.1016/j.chaos.2015.11.023. 3, 21, 44

[6] H. Yu, L. Guvenc, and U. Ozguner. Heavy-duty vehicle rollover detection and active roll control. *Vehicle System Dynamics*, 46(6):451–470, 2008. DOI: 10.1080/00423110701477529. 4

[7] B. Chen and H. Peng. A real-time rollover threat index for sports utility vehicle. *Proc. of the American Control Conference*, 2(2):1233–1237, 1999. DOI: 10.1109/ACC.1999.783564. 4, 34

[8] S. Yim. Design of a robust controller for rollover prevention with active suspension and differential braking. *Journal of Mechanical Science and Technology*, 26(1):213–222, 2012. DOI: 10.1007/s12206-011-0915-9. 6, 59

[9] Y. Zhang, A. Khajepour, and X. Xie. Rollover prevention for sport utility vehicles using a pulsed active rear-steering strategy. *Proc. of the Institution of Mechanical Engineers Part D: Journal of Automobile Engineering*, 230(9):1239–1253, 2016. DOI: 10.1177/0954407015605696.

[10] S. Yim. Design of a rollover prevention controller with H∞ preview control. *Journal of Institute of Control Robotics and Systems*, 24(1):42–48, 2018. DOI: 10.5302/j.icros.2018.17.0185.

[11] G. Dong, J. Chen, and N. Zhang. Study on the time lag between steering input and vehicle lateral acceleration response under different key vehicle parameters. *Applied Mechanics and Materials*, 226(11):681–684, 2012. DOI: 10.4028/www.scientific.net/amm.226-228.681.

[12] T. Zhu and C. Zong. Rollover warning system of heavy duty vehicle based on improved TTR algorithm. *Journal of Mechanical Engineering*, 47(10):88–94, 2011. DOI: 10.3901/jme.2011.10.088. 6

[13] Z. Jin, J. Weng, and H. Hu. Rollover stability of a vehicle during critical driving manoeuvres. *Proc. of the Institution of Mechanical Engineers Part D: Journal of Automobile Engineering*, 221(9):1041–1049, 2007. DOI: 10.1243/09544070jauto343. 6, 23, 28

[14] H. Imine, A. Benallegue, T. Madani, et al. Rollover risk prediction of heavy vehicle using high order sliding mode observer: experimental results. *IEEE Transactions on Vehicular Technology*, 63(6):2533–2543, 2014. DOI: 10.1109/tvt.2013.2292998. 9

[15] Z. Jin, L. Zhang, J. Zhang, et al. Stability and optimised H∞ control of tripped and untripped vehicle rollover. *Vehicle System Dynamics*, 54(10):1405–1427, 2016. DOI: 10.1080/00423114.2016.1205750. 9, 30, 41, 67

[16] M. Alberding, J. Tjønnås, et al. Integration of vehicle yaw stabilization and rollover prevention through nonlinear hierarchical control allocation. *Vehicle System Dynamics*, 52(12):1607–1621, 2014. DOI: 10.1080/00423114.2014.952643. 16

[17] M. Alberding, J. Tjønnas, et al. Nonlinear hierarchical control allocation for vehicle yaw stabilization and rollover prevention. *Control Conference*, 2015. DOI: 10.23919/ecc.2009.7075064.

[18] A. Kordani and A. Molan. The effect of combined horizontal curve and longitudinal grade on side friction factors. *KSCE Journal of Civil Engineering*, 19(1):303–310, 2015. DOI: 10.1007/s12205-013-0453-3.

[19] B. Han and J. Seo. Analysis of vehicle rollover using multibody dynamics. *Journal of Mechanical Science and Technology*, 30(2):797–802, 2016. DOI: 10.1007/s12206-016-0122-9.

[20] S. Zhu and Y. He. A driver-adaptive stability control strategy for sport utility vehicles. *Vehicle System Dynamics*, 55(8):1206–1240, 2017. DOI: 10.1080/00423114.2017.1308521. 16, 55

[21] W. Bao and S. Hu. Vehicle rollover simulation analysis considering road excitation. *Transactions of the Chinese Society of Agricultural Engineering*, 31(2):59–65, 2015. DOI: 10.3969/j.issn.1002-6819.2015.02.009. 17

[22] D. Baker, R. Bushman, and C. Berthelot. The effectiveness of truck rollover warning systems. *Transportation Research Record: Journal of the Transportation Research Board*, 2000. DOI: 10.3141/1779-18. 21

[23] N. Zhang, G. Dong, and H. Du. Investigation into untripped rollover of light vehicles in the modified fishhook and the sine maneuvers. Part I: Vehicle modelling, roll and yaw instability. *Vehicle System Dynamics*, 46(4):271–293, 2008. DOI: 10.1080/00423110701344752. 21

[24] R. Huston and F. Kelly. Another look at the static stability factor (SSF) in predicting vehicle rollover. *International Journal of Crashworthiness*, 19(6):567–575, 2014. DOI: 10.1080/13588265.2014.919730. 22

[25] Z. Yao, G. Wang, X. Li, et al. Dynamic simulation for the rollover stability performances of articulated vehicles. *Proc. of the Institution of Mechanical Engineers, Part D: Journal of Automobile Engineering*, 228(7):771–783, 2014. DOI: 10.1177/0954407013501486. 28

[26] D. Denis, B. Thuilot, and R. Lenain. Online adaptive observer for rollover avoidance of reconfigurable agricultural vehicles. *Computers and Electronics in Agriculture*, 126(1):32–43, 2016. DOI: 10.1016/j.compag.2016.04.030.

[27] H. Imine and M. Djemaï. Switched control for reducing impact of vertical forces on road and heavy-vehicle rollover avoidance. *IEEE Transactions on Vehicular Technology*, 65(6):4044–4052, 2016. DOI: 10.1109/tvt.2015.2470090.

[28] M. Saeedi, R. Kazemi, and S. Azadi. Improvement in the rollover stability of a liquid-carrying articulated vehicle via a new robust controller. *Journal of Automobile Engineering*, 231(3):322–346, 2017. DOI: 10.1177/0954407016639204. 28, 49, 67

[29] S. Solmaz, M. Corless, and R. Shorten. A methodology for the design of robust rollover prevention controllers for automotive vehicles. *Part 1: Differential Braking, IEEE Conference on Decision and Control*, 80(11):1739–1744, 2007. DOI: 10.1109/cdc.2006.377179. 30, 55

[30] M. Akar and A. Dere. A switching rollover controller coupled with closed-loop adaptive vehicle parameter identification. *IEEE Transactions on Intelligent Transportation Systems*, 15(4):1579–1585, 2014. DOI: 10.1109/tits.2014.2301721.

[31] H. Dahmani, M. Chadli, A. Rabhi, et al. Detection of impending vehicle rollover with road bank angle consideration using a robust fuzzy observer. *International Journal of Automation and Computing*, 12(1):93–101, 2015. DOI: 10.1007/s11633-014-0836-z. 34, 44, 67

[32] R. Azim, F. Malik, and W. Syed. Rollover mitigation controller development for three-wheeled vehicle using active front steering. *Mathematical Problems in Engineering*, article ID:918429:1–9, 2015. DOI: 10.1155/2015/918429. 51

[33] D. Chu, X. Lu, C. Wu, et al. Smooth sliding mode control for vehicle rollover prevention using active antiroll suspension. *Mathematical Problems in Engineering*, article ID:478071:1–8, 2015. DOI: 10.1155/2015/478071. 50, 86

[34] B. Mashadi, M. Mokhtari-Alehashem, and H. Mostaghimi. Active vehicle rollover control using a gyroscopic device. *Proc. of the Institution of Mechanical Engineers, Part D: Journal of Automobile Engineering*, 230(14):1958–1971, 2016. DOI: 10.1177/0954407016641322.

[35] M. Ghazali, M. Durali, and H. Salarieh. Path-following in model predictive rollover prevention using front steering and braking. *Vehicle System Dynamics*, 55(1):121–148, 2017. DOI: 10.1080/00423114.2016.1246741. 74

[36] M. Ataei, A. Khajepour, and S. Jeon. Rollover stabilities of three-wheeled vehicles including road configuration effects. *Proc. of the Institution of Mechanical Engineers, Part D: Journal of Automobile Engineering*, 231(7):859–871, 2017. DOI: 10.1177/0954407017695007. 30, 44

[37] C. Larish, D. Piyabongkarn, V. Tsourapas, et al. A new predictive lateral load transfer ratio for rollover prevention systems. *IEEE Transactions on Vehicular Technology*, 62(7):2928–2936, 2013. DOI: 10.1109/tvt.2013.2252930. 34, 36, 37

[38] H. Li, Y. Zhao, H. Wang, et al. Design of an improved predictive LTR for rollover warning systems. *Journal of the Brazilian Society of Mechanical Sciences and Engineering*, 39(10):3779–3791, 2017. DOI: 10.1007/s40430-017-0796-7. 34

[39] B. Zhu, Q. Piao, J. Zhao, et al. Integrated chassis control for vehicle rollover prevention with neural network time-to-rollover warning metrics. *Advances in Mechanical Engineering*, 8(2):1–13, 2016. DOI: 10.1177/1687814016632679. 34

[40] H. Dahmani, M. Chadli, A. Rabhi, et al. Vehicle dynamic estimation with road bank angle consideration for rollover detection: Theoretical and experimental studies. *Vehicle System Dynamics*, 51(12):1853–1871, 2013. DOI: 10.1080/00423114.2013.839819. 34, 44

[41] H. Dahmani, O. Pagès, A. Hajjaji, et al. Observer-based robust control of vehicle dynamics for rollover mitigation in critical situations. *IEEE Transactions on Intelligent Transportation Systems*, 15(1):274–284, 2014. DOI: 10.1109/tits.2013.2281135. 44, 67

[42] S. Choi. Practical vehicle rollover avoidance control using energy method. *Vehicle System Dynamics*, 46(4):323–337, 2008. DOI: 10.1080/00423110701377109. 46

[43] D. Sampson and D. Cebon. Active roll control of single unit heavy road vehicles. *Vehicle System Dynamics*, 40(4):229–270, 2003. DOI: 10.1076/vesd.40.2.229.16540. 49

[44] V. Vu, O. Sename, L. Dugard, et al. Enhancing roll stability of heavy vehicle by LQR active anti-roll bar control using electronic servo-valve hydraulic actuators. *Vehicle System Dynamics*, 55(9):1405–1429, 2017. DOI: 10.1080/00423114.2017.1317822. 49, 83

[45] V. Muniandy, P. Samin, and H. Jamaluddin. Application of a self-tuning fuzzy PI-PD controller in an active anti-roll bar system for a passenger car. *Vehicle System Dynamics*, 53(11):1641–1666, 2015. DOI: 10.1080/00423114.2015.1073336. 49, 61

[46] V. Vu, O. Sename, L. Dugard, et al. H∞ active anti-roll bar control to prevent rollover of heavy vehicles: A robustness analysis. *IFAC-PapersOnLine*, 49(9):99–104, 2016. DOI: 10.1016/j.ifacol.2016.07.503. 49, 67

[47] F. Sarel, V. Westhuizen, and P. Els. Slow active suspension control for rollover prevention. *Journal of Terramechanics*, 50(1):29–36, 2013. DOI: 10.1016/j.jterra.2012.10.001. 50

[48] Q. Zhu and B. Ayalew. Predictive roll, handling and ride control of vehicles via active suspensions. *American Control Conference*, 2014. DOI: 10.1109/acc.2014.6859037. 50

[49] W. Sun, Y. Li, J. Huang, et al. Efficiency improvement of vehicle active suspension based on multi-objective integrated optimization. *Journal of Vibration and Control*, 23(4): 539–554, 2017. DOI: 10.1177/1077546315581731. 51

[50] S. Yim, J. Choi, and K. Yi. Coordinated control of hybrid 4WD vehicles for enhanced maneuverability and lateral stability. *IEEE Transactions on Vehicular Technology*, 61(4):1946–1950, 2012. DOI: 10.1109/tvt.2012.2188921. 51

[51] N. Elmi, A. Ohadi, and B. Samadi. Active front-steering control of a sport utility vehicle using a robust linear quadratic regulator method, with emphasis on the roll dynamics. *Proc. of the Institution of Mechanical Engineers Part D: Journal of Automobile Engineering*, 227(12):1636–1649, 2013. DOI: 10.1177/0954407013502319.

[52] S. Yim. Unified chassis control with electronic stability control and active front steering for under-steer prevention. *International Journal of Automotive Technology*, 16(5):775–782, 2015. DOI: 10.1007/s12239-015-0078-2. 51

[53] B. Zhang, A. Khajepour, and A. Goodarzi. Vehicle yaw stability control using active rear steering: Development and experimental validation. *Journal of Multi-Body Dynamics*, 231(2):333–345, 2017. DOI: 10.1177/1464419316670757. 53

[54] E. Ono, Y. Hattori, Y. Muragishi, et al. Vehicle dynamics integrated control for four-wheel-distributed steering and four-wheel-distributed traction/braking systems. *Vehicle System Dynamics*, 44(2):139–151, 2006. DOI: 10.1080/00423110500385790. 54

[55] S. Solmaz, M. Akar, and R. Shorten. Adaptive rollover prevention for automotive vehicles with differential braking. *IFAC Proceedings Volumes*, 41(2):4695–4700, 2008. DOI: 10.3182/20080706-5-kr-1001.00790. 55

[56] J. Yong, F. Gao, N. Ding, et al. Design and validation of an electro-hydraulic brake system using hardware-in-the-loop real-time simulation. *International Journal of Automotive Technology*, 18(4):603–612, 2017. DOI: 10.1007/s12239-017-0060-2. 57

[57] S. Yim, K. Jeon, and K. Yi. An investigation into vehicle rollover prevention by coordinated control of active anti-roll bar and electronic stability program. *International Journal of Control Automation and Systems*, 10(2):275–287, 2012. DOI: 10.1007/s12555-012-0208-9. 59

[58] M. Doumiati, O. Senamea, L. Dugard, et al. Integrated vehicle dynamics control via coordination of active front steering and rear braking. *European Journal of Control*, 19(2):121–143, 2013. DOI: 10.1016/j.ejcon.2013.03.004. 59

[59] J. Yoon, W. Cho, J. Kang, et al. Design and evaluation of a unified chassis control system for rollover prevention and vehicle stability improvement on a virtual test track. *Control Engineering Practice*, 18(6):585–597, 2010. DOI: 10.1016/j.conengprac.2010.02.012. 59

[60] W. Liu, H. He, F. Sun, et al. Integrated chassis control for a three-axle electric bus with distributed driving motors and active rear steering system. *Vehicle System Dynamics*, 55(5):601–625, 2017. DOI: 10.1080/00423114.2016.1267368. 59

[61] M. Licea and I. Cervantes. Robust indirect-defined envelope control for rollover and lateral skid prevention. *Control Engineering Practice*, 61(4):149–162, 2017. DOI: 10.1016/j.conengprac.2017.02.008. 67

[62] R. Tafner, M. Reichhartinger, and M. Horn. Robust online roll dynamics identification of a vehicle using sliding mode concepts. *Control Engineering Practice*, 29(8):235–246, 2014. DOI: 10.1016/j.conengprac.2014.03.002.

[63] M. Licea and I. Cervantes. Robust indirect-defined envelope control for rollover and lateral skid prevention. *Control Engineering Practice*, 61:149–162, 2017. DOI: 10.1016/j.conengprac.2017.02.008. 67

[64] F. Yakub, S. Lee, and Y. Mori. Comparative study of MPC and LQC with disturbance rejection control for heavy vehicle rollover prevention in an inclement environment. *Journal of Mechanical Science and Technology*, 30(8):3835–3845, 2016. DOI: 10.1007/s12206-016-0747-8. 74

[65] M. Ghazali, M. Durali, and H. Salarieh. Vehicle trajectory challenge in predictive active steering rollover prevention. *International Journal of Automotive Technology*, 18(3):511–521, 2017. DOI: 10.1007/s12239-017-0051-3. 74

[66] B. Johansson and M. Gafvert. Untripped SUV rollover detection and prevention. *Proc. of the IEEE Conference on Decision and Control*, 5(2):5461–5466, 2004. DOI: 10.1109/cdc.2004.1429677. 83

[67] Y. Pourasad, M. Mahmoodi-k, and M. Oveisi. Design of an optimal active stabilizer mechanism for enhancing vehicle rolling resistance. *Journal of Central South University*, 23(5):1142–1151, 2016. DOI: 10.1007/s11771-016-0364-9.

[68] F. Vinicius, D. Poggetto, and A. Serpa. Vehicle rollover avoidance by application of gain-scheduled LQR controllers using state observers. *Vehicle System Dynamics*, 54(2):1–19, 2016. DOI: 10.1080/00423114.2015.1125005. 83

[69] S. Yim, C. Lim, and Myung-Hwan. An investigation into the structures of linear quadratic controllers for vehicle rollover prevention. *Proc. of the Institution of Mechanical Engineers Part D: Journal of Automobile Engineering*, 227(4):472–480, 2013. DOI: 10.1177/0954407012459462. 83

[70] H. Imine, L. Fridman, and T. Madani. Steering control for rollover avoidance of heavy vehicles. *IEEE Transactions on Vehicular Technology*, 61(8):3499–3509, 2012. DOI: 10.1109/tvt.2012.2206837. 86

Authors' Biographies

ZHILIN JIN

Zhilin Jin is currently an Associate Professor at Nanjing University of Aeronautics and Astronautics, focusing on vehicle system dynamics and control and by-wire systems for intelligent connected vehicles. He received his Ph.D. in Vehicle Engineering from Nanjing University of Aeronautics and Astronautics, Nanjing, China in 2008. Dr. Jin has more than 17 years of research experience in and has written 40 papers on vehicle rollover dynamics, vehicle rollover warning, vehicle rollover prevention, electro hydraulic braking system, and steer by-wire system. He has been a reviewer of many international journals, including *Vehicle System Dynamics*, *IEEE Transactions on Vehicle Technology*, *IEEE Transactions on Human-Machine Systems*, *Mechatronics*, *International Journal of Heavy Vehicle Systems*, *IET Intelligent Transport Systems*, and so on.

BIN LI

Bin Li is currently with Aptiv PLC (USA), focusing on autonomous driving algorithm development and verification of motion planning and vehicle control. He received his Ph.D. in Mechanical Engineering from Shanghai Jiao Tong University, Shanghai, China in 2010. Dr. Li has more than 15 years' research experience in vehicle dynamics and control, electric vehicles, active safety, and autonomous driving, and has written over 50 papers and chapters. He was Guest Editor of the *International Journal of Heavy Vehicle Systems*, *International Journal of Vehicle Autonomous Systems*, and *IET Intelligent Transport Systems*. Dr. Li currently serves as Associate Editor of the *International Journal of Vehicle Autonomous Systems* (IJVAS), *SAE International Journal of Commercial Vehicles*, and *SAE International Journal of Passenger Cars: Electronic and Electrical Systems*. He has been an active organizer for SAE World Congress and ASME conferences since 2015.

JINGXUAN LI

Jingxuan Li is currently a graduate student at Nanjing University of Aeronautics and Astronautics, focusing on vehicle system dynamics and control and by-wire system for intelligent connected vehicles. He received his Bachelor's degree in Vehicle Engineering from Nanjing University of Aeronautics and Astronautics, Nanjing, China in 2018.

Printed in the United States
by Baker & Taylor Publisher Services